The Meat Crisis

Meat debates:
List of Seminars:

Add notes to ~~term~~ reflections
as you read each chapter.

This book is dedicated to the memory of Peter Roberts, farmer, visionary and founder of Compassion in World Farming.

The Meat Crisis

Developing More Sustainable Production and Consumption

Edited by
Joyce D'Silva and John Webster

To Ruth and Peter with lots of love Joy & xt

earthscan

publishing for a sustainable future

London • Washington, DC

First published in 2010 by Earthscan

Earthscan Ltd, Dunstan House, 14a St Cross Street, London EC1N 8XA, UK
Earthscan LLC, 1616 P Street, NW, Washington, DC 20036, USA
Earthscan publishes in association with the International Institute for Environment and Development

For more information on Earthscan publications, see www.earthscan.co.uk or write to earthinfo@earthscan.co.uk

ISBN: 978-1-84407-902-5 hardback
ISBN: 978-1-84407-903-2 paperback

Typeset by Composition and Design Services
Cover design by Susanne Harris and Elli Goodlet

A catalogue record for this book is available from the British Library

Library of Congress Cataloging-in-Publication Data

The meat crisis : developing more sustainable production and consumption / edited by Joyce D'Silva and John Webster. – 1st ed.
    p. cm.
   Includes bibliographical references and index.
   ISBN 978-1-84407-902-5 (hardback) – ISBN 978-1-84407-903-2 (pbk.) 1. Animal industry–Environmental aspects. 2. Meat industry and trade–Environmental aspects. 3. Sustainable agriculture. 4. Animal culture. 5. Livestock. I. D'Silva, Joyce. II. Webster, John, 1938-
   HD9410.5.M43 2010
   338.1'76–dc22

2010015250

At Earthscan we strive to minimize our environmental impacts and carbon footprint through reducing waste, recycling and offsetting our CO_2 emissions, including those created through publication of this book. For more details of our environmental policy, see www.earthscan.co.uk.

FSC
Mixed Sources
Product group from well-managed
forests and other controlled sources
Cert no. SGS-COC-2482
www.fsc.org
© 1996 Forest Stewardship Council

Printed and bound in the UK by TJ International
The paper used is FSC certified and the inks are vegetable based.

Contents

List of Figures and Tables

Figures

Tables

List of Contributors

Ainslie Butler, a graduate in biomedical sciences, is currently working towards a PhD at the Australian National University where she is studying the impacts of climate on food-borne disease in Australia. She has also recently contributed to a study investigating the co-benefits of reduced meat consumption on population health, as part of a larger Wellcome Trust-funded research programme exploring the health co-benefits of various climate change mitigation activities, resulting in a series of articles published in *The Lancet* in late 2009.

Dr Andy Butterworth is a veterinarian who leans towards animal welfare science and ethics, who worked first in practice, and then in the area of applied animal welfare work, particularly in poultry, dairy cattle and the welfare of fur-bearing and marine mammals. Recently, Andy was coordinator of part of a large European initiative to create welfare assessment protocols for farm animals in Europe called Welfare Quality. Andy has presented his work on animal welfare topics in many international academic and 'legislative' fora including EU Commission, UK Government, International Whaling Commission, Asian fora on poultry issues, and has been part of animal welfare research and training in Asia, north and south America, Australasia and Europe.

Martin Caraher is Professor of Food and Health Policy in the Centre for Food Policy at City University, London, UK. He has worked on issues related to food poverty, cooking skills, local sustainable food supplies, the role of markets and cooperatives in promoting health, farmers markets, food deserts and food access, retail concentration and globalization.

Joyce D'Silva is the Director of Public Affairs and former Chief Executive of Compassion in World Farming, the leading charity advancing the welfare of farm animals worldwide through research, education and advocacy. Joyce lectures internationally and has published widely on the welfare of farm animals, including co-editing *Animals, Ethics and Trade* (Earthscan, 2006); 'Sustainable Agriculture' (2008) in a report by Climate Caucus to the United Nations Secretary General; 'Modern farming practices and animal welfare' in *Ethics, Law and Society* Volume IV (2008); 'The urgency of change' in *The Future of Animal Farming* (Blackwell, 2008); and *Policies for Sustainable Production and Consumption* (Springer, in press) for the proceedings of the 2009 conference, 'The Integration of Sustainable Agriculture and Rural Development in the Context of Climate Change, the Energy Crisis and Food Insecurity'.

Ian J. H. Duncan is Professor Emeritus at the Department of Animal and Poultry Science, University of Guelph, Canada. An expert in applied ethology, he started his career at the University of Edinburgh. He has been given numerous awards for his work. To date he has published one edited book, 31 book chapters, 106 papers in refereed scientific journals, 77 papers in refereed conference proceedings, 150 papers in other conference proceedings and many abstracts in scientific journals.

Tara Garnett is a researcher at the University of Surrey. Her work focuses on researching the contribution that our food consumption makes to UK greenhouse gas emissions and the scope for emissions reduction, looking at the technological options for GHG reductions, at what could be achieved by changes in behaviour and how policies could help promote both these objectives. She is also interested in the relationship between emissions reduction objectives and other social and ethical concerns, including human health, animal welfare and biodiversity. In addition, Tara runs The Food Climate Research Network. This brings together a growing number of individuals from across the food industry, non-governmental, government and academic sectors and from a broad variety of disciplines to share information on issues relating to food and climate change.

Michael Greger is a graduate of the Cornell University School of Agriculture and the Tufts University School of Medicine. He serves as Director of Public Health and Animal Agriculture at Humane Society International. His recent scientific publications in *American Journal of Preventive Medicine, Biosecurity and Bioterrorism, Critical Reviews in Microbiology*, and the *International Journal of Food Safety, Nutrition, and Public Health* explore the public health implications of industrialized animal agriculture, and his latest book, the acclaimed *Bird Flu: A Virus of Our Own Hatching*, is now available full-text at no cost at www.BirdFluBook.org.

Susanne Gura has been studying the livestock and aquaculture industries for several years. She has done analytical and advocacy work for Econexus,

Greenpeace, the League for Pastoral Peoples and others, and for many years has advised the German government on international agricultural research policies. She holds academic degrees in human nutrition and rural development.

Dr Fredrik Hedenus is a researcher at the Department of Physical Resource Theory, Chalmers University of Technology, Sweden. He has published scientific papers in the areas of greenhouse gas emissions from food, energy system modelling, energy security and technological change.

Arjen Y. Hoekstra is professor in water resources management at the University of Twente, Enschede, The Netherlands. Hoekstra is the creator of the water footprint concept. His publications include the books *Perspectives on Water* (International Books, 1998) and *Globalization of Water* (Wiley-Blackwell, 2008).

Tim Lang is Professor of Food Policy at City University, London, UK and is a member of the Sustainable Development Commission. He is a regular advisor to the World Health Organization at global and European levels. He has been a special advisor to four House of Commons Select Committee inquiries and an advisor to the Foresight Obesity programme. He is a member of the Royal Institute of International Affairs (Chatham House) 'Food Supply in the 21st Century' working party and team. He has authored and co-authored over 120 publications.

Professor Alistair Lawrence studied for his PhD under Professor David Wood-Gush and then took on responsibility for behaviour and welfare research at the Scottish Agricultural College (SAC). Currently, he heads the Animal Welfare Team at SAC and is responsible for coordinating Scottish Government research into livestock welfare. He also has an interest in developing wider public understanding of animal welfare particularly in young people. Recently he took on the role of acting head of the Sustainable Livestock Group in SAC. He holds a joint position with the University of Edinburgh Veterinary School where he helps oversee delivery of welfare teaching to undergraduate veterinary and masters students. He recently finished a nine-year spell as a member of the UK Farm Animal Welfare Council.

Anthony J. McMichael, epidemiologist, is Professor of Population Health at the Australian National University, Canberra, where he heads the research programme on Environment, Climate and Health (at the national Centre for Epidemiology and Population Health). He has conducted wide-ranging research on food, nutrition and disease, and, more recently, has played a central role in the work of the UN's Intergovernmental Panel on Climate Change (IPCC) and the World Health Organization in the assessment of health risks from climate change. His books include *Human Frontiers, Environments and Disease: Past Patterns, Uncertain Futures* (Cambridge University Press, 2001) and (with co-authors) *Climate Change and Human Health: Risks and Responses* (WHO/UNEP, 2003).

Dr Rajendra Pachauri is Chairman of the Intergovernmental Panel on Climate Change, which won the 2007 Nobel Peace Prize for work on climate change (shared with Al Gore). He is also the Director General of The Energy and Resources Institute (TERI) in India. To acknowledge his immense contribution in the environmental field, he has been awarded the Padma Bhushan, one of India's highest civilian awards. He holds doctorates in Industrial Engineering and Economics and has lectured in the US and in India and has been a Visiting Research Fellow at the World Bank. He has advised the United Nations, the governments of Japan and India and numerous international organizations.

Martin Palmer is Secretary General of the Alliance of Religions and Conservation and co-chair of the UN Programme on Faiths, Climate Change and the Natural Environment. He is author of over 20 books as well as a number of translations of Chinese classics and regularly presents for both BBC TV and radio on religious, cultural and environmental issues.

Jonathon Porritt, CBE, Co-Founder of Forum for the Future, is an eminent writer, broadcaster and commentator on sustainable development. He was the former long-serving Chair of the UK Sustainable Development Commission, is Co-Director of The Prince of Wales's Business and Environment Programme, a Trustee of the Ashden Awards for Sustainable Energy and is involved in the work of many non-governmental organizations (NGOs) and charities as Patron, Chair or Special Adviser. His latest books are *Capitalism as if the World Matters* (Earthscan, revised 2007), *Globalism & Regionalism* (Black Dog, 2008) and *Living Within Our Means* (Forum for the Future, 2009). Jonathon received a CBE in January 2000 for services to environmental protection.

Dr Kate Rawles lectured in environmental ethics at Lancaster University (UK) before going freelance. In 2002 she won a NESTA Fellowship to develop Outdoor Philosophy and now works part-time as a senior lecturer in Outdoor Studies at the University of Cumbria in between running Outdoor Philosophy courses. Kate lectures and writes on a range of environmental and animal welfare issues, with a focus on values and sustainable development. She is a member of the Food Ethics Council, the academic director of Forum for the Future's innovative 'Reconnections' course and a board member and lecturer on Forum's MSc in Leadership for Sustainable Development. In 2006, Kate cycled from Texas to Alaska, exploring North American attitudes to climate change. She is currently finishing *The Carbon Cycle – Crossing the Great Divide* based on this journey. She is a member of the Food Ethics Council and a Fellow of the Royal Geographical Society.

Mike Rayner is Director of the British Heart Foundation (BHF) Health Promotion Research Group, which is based within the Department of Public Health of the University of Oxford. The Group carries out research into food advertising and labelling, so-called 'fat taxes' and the relationship between a healthy diet and a sustainable diet. Mike is Vice Chair of Sustain, the alliance for better food and farming and a trustee of the National Heart Forum in the UK. He is Chair of the Nutrition Expert Group for the European Heart Network based in Brussels. Mike is also Assistant Curate, St Matthew's Church, Oxford.

Peter Scarborough is a senior researcher within the BHF Health Promotion Research Group at the University of Oxford. He is the lead researcher on a programme of research aimed at describing the burden of cardiovascular disease and associated risk factors in the United Kingdom. He is also involved in work on the links between public health and environmental sustainability and developing and validating nutrient profile models that provide a definition of unhealthy foods. Pete studied mathematics at undergraduate level and retains an interest in statistics and mathematical modelling.

Alistair Stott is Reader in Animal Health Economics at the Scottish Agricultural College (SAC) and based in Edinburgh, UK. He heads a team of researchers in bio-economics and rural strategy within SAC's Land Economy and Environment Research Group. His team's research is focused on the multiple roles of modern farming, the trade-offs this entails and the science, technology and policy required to reconcile them.

Colin Tudge is a biologist by education and a writer by trade, mainly of books on natural history and agriculture but with a keen interest too in politics and the relationship between science and religion. In 2009, together with his wife,

Ruth, he established the Campaign for Real Farming (www.campaignfor-realfarming.org). His latest published books include *So Shall We Reap* (Allen Lane, London, 2003), *The Secret Life of Trees* (Allen Lane, London, 2005), *Feeding People is Easy* (Pari Publishing, Italy, 2007) and *Consider the Birds* (also known as *The Secret Life of Birds*; Penguin, London, 2009).

John Webster is Professor Emeritus at the University of Bristol, UK. On arrival at Bristol in 1977 he established a unit for the study of animal behaviour and welfare, which is now over 50 strong. He was a founder member of the Farm Animal Welfare Council and first propounded the 'Five Freedoms', which have gained international recognition as standards for defining the elements of good welfare in domestic animals. He is a former President of both the Nutrition Society and the British Society for Animal Science. His book *Animal Welfare: A Cool Eye towards Eden* was published in 1994 (Wiley-Blackwell, London) and is still in print. Its successor, *Animal Welfare: Limping towards Eden* appeared in 2005.

Dr Stefan Wirsenius received a PhD in Environmental Sciences at Chalmers University of Technology, Sweden. He is currently is a researcher in the Department of Energy and Environment, at Chalmers University of Technology.

Michelle Wu is a Research Assistant at the Centre for Food Policy, City University, London, UK. She has a background in nutrition and public health and has worked in both the UK and Canada in these capacities. Recent work includes evaluations of initiatives to increase healthy and sustainable food and improve food skills in public sector catering in London, school food training for dinner ladies and a review of cooking and young people.

Richard Young has farmed organically in England since 1974. He played a leading role in the development of organic standards and certification and is currently a policy adviser to the Soil Association. He is a past editor of *New Farmer & Grower* (now *Organic Farming*) and has written a series of reports for the Soil Association on 'The use and misuse of antibiotics in UK agriculture' and other livestock farming issues. He farms 400 acres in the Cotswolds with his mother and sister.

Acknowledgements

I would like to thank John Webster for his brilliant editorial work. Not much escapes his academic eagle eye! As ever, he was a delight to work with. I am so grateful to the chapter authors for the time and energy they have given to their input. Together they have helped create a wealth of knowledge and creative thinking. Finally, I would like to thank my colleague at Compassion in World Farming, Wendy Smith, who has been a first rate editorial assistant and support.

Joyce D'Silva

Foreword

I am delighted to be able to write the foreword for this book, *The Meat Crisis: Developing More Sustainable Production and Consumption*, which has valuable contributions from a number of distinguished authors. I am very happy that Earthscan, a publisher that evokes a great deal of respect worldwide, has produced this extremely timely volume. There are, of course, several aspects of the meat crisis that deserve attention and which have been covered explicitly in the book, but my current focus is on the implications of the entire meat cycle for emissions of greenhouse gases, which are increasing rapidly as a result of higher meat consumption worldwide. As it happens, several societies that in the past had very low levels of meat consumption are now demanding larger quantities as a result of increased incomes and very effective marketing strategies pursued by the industry.

I have been reflecting personally on this issue for several years, particularly after I became a vegetarian myself in the realization that consuming meat was neither necessary for the good health of a human being, nor was it desirable if one was concerned about protecting the environment. However, vegetarianism was a personal preference, and I did not wish to speak about it in public. However, then in mid 2007 I was in Bangkok at the end of the plenary session of the Intergovernmental Panel on Climate Change (IPCC) where the Working Group III report, as part of the Fourth Assessment Report (AR4), was approved and released. At the end of that session a press conference was held to inform members of the media about the main findings of the report, covering a detailed assessment of mitigation measures to reduce the emissions of greenhouse gases, which are responsible for human-induced climate change. For the first time in that report we had highlighted the role of lifestyle changes as part of mitigation actions. One of the newspersons present on the occasion asked me a question regarding what kinds of lifestyle changes would reduce emissions of greenhouse gases? I responded by stating that a reduction in meat consumption would certainly be an important change in lifestyle. I had done some analysis and study of the meat cycle in arriving at this view. Having made this statement public, which did get coverage in several parts of the world, I felt that this was an issue that needed the attention of the global community, because otherwise all the economic and demographic changes that were taking place round the world will result in increased emissions of greenhouse gases from this sector. I am delighted to observe that wherever I have spoken on the subject or written about it I have received a very enthusiastic response. Of course, there are some who ridicule the results I present and the conclusions I offer, because some believe that the concentration of greenhouse gases in the atmosphere has nothing to do with climate change and others completely ignore their role in being responsible for emissions from the meat cycle. But as I have always conveyed in my speeches, 'if you eat less meat you would be healthier and so would the planet'. We really need to ensure that we keep both aspects of this statement in mind and while perhaps the ideal situation would be for human beings to avoid meat consumption completely, for a start we should give up red meat and perhaps consume lesser quantities of poultry products as well. I would not suggest any ordinance or laws in this direction because dietary preference is very much a matter of individual choice and we only need to appeal to reason and logic to ensure that people understand the benefits of lifestyle changes that they follow by reducing dependence on diets consisting of animal protein.

I am sure this book will create a major influence through wide readership all over the world and I hope people will act on the basis of the knowledge that is conveyed through its pages.

R. K. Pachauri

Director General, The Energy and Resources Institute (TERI) and Chairman, Intergovernmental Panel on Climate Change (IPCC)

List of Acronyms and Abbreviations

ADF	Assured Dairy Farms
AICR	American Institute for Cancer Research
ARC	Alliance of Religions and Conservation
AREN	Association for the Re-dynamization of Livestock in Niger
BAU	'business-as-usual'
BSE	Bovine Spongiform Encephalopathy
CAFO	Confined Animal Feeding Operation
CAP	Common Agricultural Policy
CEH	Centre for Ecology and Hydrology
CH_4	methane
CHD	Coronary Heart Disease
CIA	Central Intelligence Agency
CO_2e	CO_2 equivalent
CSIRO	Commonwealth Scientific and Research Organization
CVD	Cardiovascular Disease
DALY	Disability Adjusted Life Years
Defra	UK Department for the Environment, Food and Rural Affairs
DHA	docosahexaenoic acid
EASAC	European Academies Science Advisory Council
EBLEX	English Beef and Lamb Executive
EEA	European Environment Agency
EFSA	European Food Safety Authority
EU	European Union
FABRE	Farm Animal Breeding and Reproduction Technology Platform
FAO	Food and Agriculture Organization of the United Nations
FCE	Feed Conversion Efficiency
FCR	Feed Conversion Ratio
FDA	Food and Drug Administration
FSC	Forest Stewardship Council
GE	genetically engineered
GHG	greenhouse gas

GM	genetically modified
GMO	genetically modified organism
GPS	Global Positioning System
ha	hectare
HACCP	Hazards Analysis Critical Control Point
IAASTD	International Assessment of Agricultural Science and Technology Development
IPCC	Intergovernmental Panel on Climate Change
IUCN	International Union for Conservation of Nature
LCA	life cycle analysis
l/day	litres per day
LRNI	Lower Reference Nutrient Intake
LU	livestock units
MAFF	UK Ministry of Agriculture, Fisheries and Food (replaced by Defra)
MBM	meat and bone meal
MDPs	meat and dairy products
ME	metabolizable energy
MRSA	Methicillin-Resistant *Staphylococcus Aureus*
MSC	Marine Stewardship Council
mt	millions of tonnes
NCD	Non-Communicable Diseases
NDFAS	National Dairy Farm Assured Scheme
NGO	non-governmental organization
NHS	National Health Service
NINA	Norwegian Institute for Nature Research
NO_2	nitrous oxide
OECD	Organisation for Economic Co-operation and Development
OIE	Office International des Epizooties
PE	Partial Equilibrium
PEF	Production Efficiency Factor
PMQ	Personal Meat Quota
ppb	parts per billion
ppm	parts per million
REDD	Reducing Emissions from Deforestation and Degradation
SACN	Scientific Advisory Committee on Nutrition
SARS	Severe Acute Respiratory System
SCC	Somatic Cell Counts
SDC	Sustainable Development Commission
TERI	The Energy and Resources Institute (India)
TMR	Total Mixed Ration
USDA	US Department of Agriculture

UV	ultraviolet
WCRF	World Cancer Research Fund
WHO	World Health Organization
WTO	World Trade Organization
WWF	World Wildlife Fund

Introduction

Joyce D'Silva and John Webster

The first response to the title of this book could well be: 'What meat crisis?' Indeed supermarket shelves and butchers' shops are well stocked with every kind of meat. Special offers on packs of chicken legs abound. At one point a 1 kilo pack of chicken portions cost the same in a leading UK supermarket as a pack of four cans of beer. In fact, in the developed world, meat has probably never been so cheap.

Alongside the abundance of meat in our shops, restaurants and fast food outlets has come the increase in consumption. Some people must be eating maybe ten times as much meat per year as their great-grandparents consumed, possibly much more. Meat is no longer the special occasion food, like the fatted calf of Biblical fame; we don't have roast chicken as an expensive Sunday lunch special any more. Those on average incomes can eat meat three times a day and feel no poorer for it.

The truth is – this IS the Meat Crisis! Too much meat for our own good health, too much for the dwindling global resources of land and water, too much for the health of our planet's climate and environment and too much to enable the animals we eat to have decent lives before we devour them.

This book is not a vegetarian gospel, although the editors will have no problem if any reader, having read the facts we present, makes that choice. Rather, this book is a call to policy makers, academics and citizens to take steps to reduce meat production and consumption to sustainable levels.

The authors whom we have engaged represent key expertise in their respective fields. They each create a compelling part of the picture, and an alarming picture it is. Global meat and milk consumption is expected to roughly double in the first half of the 21st century. That means, barring some increases in 'efficiency' of production, an approximate doubling of the number of farm animals we slaughter for our consumption each year – from around 60 billion to maybe 120 billion.

A huge amount of epidemiological research is being amassed, showing connections between over-consumption of meat with, for example, colon cancer. The saturated fats so abundant in most meats are recognized as contributing to the growing global epidemic of obesity and its knock-on conditions such as Type-2 diabetes and heart disease. As Tony McMichael and Ainslie Butler (Chapter 11) and Mike Rayner and Peter Scarborough (Chapter 12) point out, whilst recognizing the nutritional value of some meat constituents in the diet and its contribution, in small amounts, to raising nutritional levels in the malnourished, it is its over-consumption, so easily achieved in our cheap-meat era, which concerns the health experts – and should concern all consumers. We should also not forget the associated threat to human health posed by diseases from livestock being passed to humans, a scenario ably addressed by Michael Greger in Chapter 10.

As Tara Garnett shows in Chapter 3, livestock production is responsible for nearly one fifth of the greenhouse gas emissions that result from human activity – more than transport globally. In fact most of us could do more for the climate by cutting our meat and dairy consumption rather than our car and plane journeys – although the editors would advocate choosing to do both!

We know that the Earth's forests help keep our planet in good health – and support an astonishing range of wild creatures and plants. Yet their devastation in South America has been mainly for cattle ranching and the growing of soya to feed primarily to farmed animals located in other parts of the world. Susanne Gura (Chapter 4) catalogues the impact of industrial farming on biodiversity. Our best efforts to stop the destruction of our forests appear as yet to be no match for the bottomless pit of intensive livestock production (in Indonesia, our appetite for palm oil as a ubiquitous ingredient in a myriad of foods is having a similar effect).

With around 94 per cent of global soya production destined, not for tofu-eating vegans, but for cattle, pigs and poultry, and with nearly 40 per cent of cereals also destined for livestock consumption, we need only do the sums to see what an inefficient use of precious resources this is. Animals consume far more in calories than they can yield for our consumption. Meanwhile much of that land could be growing crops for direct human consumption. Already, the new trend for water and land-scarce countries to buy up vast tracts of cheap land elsewhere in the world has been undertaken, not just to grow crops for their people, but to grow feed crops for their intensively farmed animals.

Water itself may become the crunch factor in the meat crisis – and the editors urge you to read Arjen Hoekstra's chapter (2), which should send alarm bells ringing in the corridors of power and policy making.

Excreta from billions of farm animals has to go somewhere and all too often it has leached into waterways, creating poisonous downstream pollution. Amazingly, in some countries farmers who pollute water courses are exempt from prosecution under environmental protection laws! Meanwhile the methane and nitrous oxide from animal slurry act as toxic greenhouse gases and the ammonia contributes to acid rain.

The predicted growth in farm animal numbers is unlikely to be achieved without industrial factory farming of animals. Webster, Duncan, Butterworth, and Lawrence and Stott (Chapters 6 to 9) spell out just what horrors this has meant for the animals incarcerated in such units. Truthfully we have, in many cases, selectively bred animals whose own bodies may be their worst enemy, susceptible to lameness and increasingly so vulnerable to disease that most would be unable to live out their 'natural' life span. To add insult to injury, we then confine them in cages, crates or concrete pens and sheds, where they are unable to act as their inheritance urges them: hens can no longer build nests, flap their wings or lay their eggs in privacy, pigs have no soil or other substrate to root in, many breeding sows may spend several weeks of each year unable even to turn round in their farrowing crates and natural grazers like cattle may be kept far from grassy fields, thus throwing their natural sleep-and-eat patterns into disarray and their overburdened digestive systems into uncomfortable acidic overload.

Are there solutions to the meat crisis? Of course there are, but their adoption and implementation will require vision, courage and, probably, financing in the initial stages.

The easiest solution may be for those in the developed world and middle class consumers in rapidly developing transition countries like China and Brazil to radically cut their personal meat consumption. It should not be beyond the imagination of the advertising industry to develop great slogans to assist health and environmental authorities in achieving this cost-effective action. Government expenditure on many other areas of health could ease as a result.

Conservation measures and organic farming can help maintain precious soils and biodiversity and reduce application of artificial nitrogenous fertilizers and noxious pesticides, as Richard Young illustrates in his fascinating chapter (5). Animal manures can be returned to the soil in reasonable amounts to enrich it. Some can be treated in anaerobic digesters to become a local energy source.

Although the excessive demand for meat can lead to endemic overgrazing and deforestation, a more modest consumption of higher value, higher welfare meat could support the conservation and expansion of permanent grasslands that allowed cattle to graze (rather than consume soya and cereals). Permanent grasslands sequester carbon in the soil and offset the global warming effects of the methane belched by cattle and other ruminants as a natural consequence of rumen fermentation.

Reducing meat production and overgrazing will lighten the burden on soils and waters and remove the *raison d'être* for much deforestation.

From the animal welfare viewpoint, smaller numbers of animals have the potential to be reared in more welfare-friendly conditions. Grazing land can not only sequester carbon, it can provide a good environment for ruminants to live in, provided of course that they have shelter, supplementary feed when necessary and are managed well.

Lower animal numbers could give the remaining animals more space in their indoor accommodation and lead to the end of the close confinement cage and crate systems that have destroyed the life quality of so many animals and brought the phrases 'factory farming' and 'intensive farming' into justifiable disrepute. It is vital that we put into practice, not just the fine words in the Lisbon Treaty, which recognize animals as 'sentient beings', but that we actively promote husbandry systems and practices that enable the animals to enjoy their lives.

Colin Tudge and John Webster (Chapters 1 and 6) both remind us to not to forget the farmer, who produces our food and can play a vital role in achieving a truly greener countryside and maintaining it in a way that benefits local biodiversity too. The editors have no doubt that farmers want to take pride in their work. By reducing the numbers of animals they keep, by being helped to keep them in more welfare-friendly conditions, and by diversifying where opportunity offers, we have no doubt that farmers' own emotional well-being will also be enhanced. For the few so-called farmers who have invested their lives in mass production with global agribusiness, we can only suggest they find an alternative way of farming, so that they can truly earn that honourable title of 'farmer'.

This book focuses on meat and concentrates on farmed mammals and birds. However, we are aware that many millions of people rely on fish as important sources of protein. As the oceans are increasingly depleted of wild fisheries stocks, aquaculture, including fish farming, is becoming more significant. Tara Garnett (in Chapter 3) and Susanne Gura (Chapter 4) mention some of the implications of the growing fish farming industry on climate, resource consumption and biodiversity. As science confirms for us that fish can indeed feel pain, the health and welfare of fish are also issues that must be included in our assessment, but a detailed discussion of these issues is beyond the scope of this book.

We have not forgotten that culture, religion and personal ethical value systems may all play an important role in our lives and be invoked when making dietary decisions. Kate Rawles (Chapter 13) challenges us to develop ethical farming and food and Martin Palmer (Chapter 14) ably spells out the reasoning behind some of these choices.

Compassion in World Farming and Friends of the Earth commissioned an expert report in 2009. Its findings show clearly that it will be possible to feed the predicted human population in 2050 both humanely and sustainably, but only if the major meat-eating nations reduce their consumption (Compassion in World Farming and Friends of the Earth (UK), 2009).

The greatest challenge is not to academics, who now have a wealth of research to hand, nor to individuals who are motivated to modify their diets, but to governments, global institutions and policy makers, who have to find ways to achieve the desired levels of meat production and consumption.

This is why we have included an important section on policy (Chapters 15 to 17), where leading thinkers like Stefan Wirsenius and Fredrik Hedenus, Tim Lang and colleagues, and Jonathan Porritt set out their recommendations, some of them truly radical, for the best way forward.

Finally we urge all our readers to be visionaries, to work at any level from personal to global to help bring about a greener and more compassionate world, where we can all access healthier food, where the planet's precious resources are nurtured, not devastated, where our carbon footprint is reduced and where the animals we farm for our food can have lives worth living. In this way we can have a level of meat and dairy production and consumption that is better for people, animals and our planet.

References

Compassion in World Farming and Friends of the Earth (UK) (2009) 'Eating the planet? Feeding and fuelling the world sustainably, fairly and humanely – a scoping study', available at www.ciwf.org/eatingtheplanet, last accessed 24 May 2010

Part 1

The Impacts of Animal Farming on the Environment

1

How to Raise Livestock – And How Not To

Colin Tudge

Most shocking in the modern world is the contrast between what could be and what is. Biologically speaking, the human species should be near the beginning of its evolutionary run. Tsunamis and volcanoes happen, and asteroids are a constant threat, but the long history of the world suggests that with average luck our species should last for another million years – and then our descendants might draw breath and contemplate the following million. Experience and simple extrapolation suggest that if we manage the world well, it is perfectly capable of supporting all of us to a very high standard through all that time – the 6.7 billion or so who are with us now, and the 9.5 billion who will be with us within a few decades. Yet we are being warned from all sides – not simply, these days, by the professional environmentalists – that the human species will be lucky to survive through the present century in a tolerable or even in a recognizable form; and the warnings look all too plausible.

We face what is now customarily called 'the triple crunch': the collapse of the world's banks, 'peak oil' (meaning most simply that from now on demand will exceed any possibility of supply, until the oil finally runs out altogether) and, of course, global warming, which could make nonsense of all other predictions. In truth, though, the 'crunch' is hydra-headed: there was a great deal wrong with the world even in the days when the banks seemed to be riding high, oil was assumed to be inexhaustible if not literally infinite and global warming was written off as the fantasy of a few eccentric physicists. To the recognized disasters of the triple crunch we should at least add the squandering of fresh water, the steady dissipation of phosphorus, the loss of forest and virtually all other terrestrial environments and the serious depletion of the oceans through warming, pollution and the destruction of fish nurseries including mangroves and coral reefs. Then of course there is the present plight of human beings – one billion undernourished, constantly on the edge of famine; one billion eating too much of the wrong things, with obesity just an indication of much worse (heart disease, cancers, diabetes), and for good measure (a different kind of statistic, but related), one billion, and rising, now living in urban slums. In addition, about half of all our fellow creatures are in imminent danger of extinction, though this is surely a conservative estimate.

Agriculture is at the heart of all these setbacks – affected directly by all of them, and a significant cause of most of them. And at the heart of all that is wrong with agriculture is livestock – again the victim and a principal cause of much that is awry with all the rest.

Nothing can be put right *ad hoc*. Everything depends on everything else. No individual mistake can be corrected without attending to all the others. We can't really put agriculture back on course – not at least on a secure and stable course – unless we also create an economy that is sympathetic to sound farming, instead of one that, as now, makes it well nigh impossible to farm sensibly without going bust. Farmers cannot produce good food by good means and sell it for what it is really worth unless people at large are prepared to pay for it – which means the world needs to restore its food cultures; and unless people are able to afford it – which brings us back to the economy. We cannot tackle any of the problems unless we give a damn, which is a matter of morality. It is very difficult, too, to make serious changes without the assistance or at least the compliance of governments – but the world's most powerful governments no longer seem to think it is their job to govern. They interfere with our lives but that is not the same thing at all. Certainly, Britain's governments since the 1970s have as a matter of policy handed over their traditional powers to the 'the market', and to international agencies such as the World Trade Organization, which affect to oversee that market.

It is all a great pity. The world is still a wonderful place – extraordinarily productive and obliging, and far more resilient than we have a right to expect. Human beings are intelligent, skilful and despite appearances are steeped in what Adam Smith called 'natural sympathy'; not wanting their fellows or their fellow creatures to suffer; recognizing the debt that we all owe to each other. We have science, which gives us extraordinary insights, and ought to be one of the great assets of humankind, and technologies that give us extraordinary power, including 'high' technologies – the kind that emerge, often counterintuitively, from science. We should not be in a mess. But science and technology, two of the great triumphs of humankind, alas are as likely these days to be our enemy. Scientists are identifying what is wrong with the world to be sure, and this is vital. But they are also stoking the fires.

But is all this really the case? Is it possible to feed everybody well, now and in the future? Is livestock really so important to the whole unfolding scene? The answer to all these questions is a resounding 'Yes'.

How to feed everybody well and why livestock is crucial

Farming has to be productive of course but it also has to be both sustainable and resilient: not wrecking the rest of the world as the decades pass; and able to change direction as conditions change – which is especially necessary as the climate shifts. But nature itself shows how all this can be achieved. Wild nature has been continuously productive through hugely turbulent changes of circumstance for 3.8 billion years. Farming is an artifice of course – a human creation – but if we want it to serve us well, and go on serving us, and not kill everything else, then we should seek to emulate nature.

So how does nature achieve its ends? In general, it is self-renewing – it taps in to renewable energy, which mainly (though not quite exclusively) is solar energy, and recycles the non-renewables: carbon, nitrogen, phosphorus, water and all the rest. It manages all this with enviable efficiency because it is so diverse – millions of different species (we don't know how many) acting in rivalry but also more importantly in concert; and it's because the system is so diverse that it is also so resilient. No one disaster can destroy everything – there have always been survivors through all the mass extinctions – and over time, all the creatures within the system, and the ecosystems as a whole, evolve and adapt. The relationship between plants and animals, which are often perceived to be 'dominant' because they are big, is synergistic. The plants convert solar energy and minerals into carbohydrates, proteins and the rest; and the animals help to keep everything cycling.

Farming that emulated nature would be the same: deriving its inputs from renewables; wasting nothing; diverse; balancing crops (plants) against livestock (animals). In principle this is simple. Natural.

In practice, in basic structure, this is how agriculture generally has been for the past 10,000 years. The emphasis in most traditional farming has been on crops, which must be grown on the prime land. The staples, mainly cereals, are grown by arable techniques (on the field scale), while most of the rest are raised by horticulture (on the garden scale). Traditional farmers everywhere commonly grew and grow many different crops – often a huge variety. Except in extreme circumstances (as in extensive grasslands and the highest altitudes and latitudes) livestock are kept mainly or exclusively to supplement the crops. They too are diverse – if not in species then certainly genetically. Overall, traditional farming is and was 'polycultural'.

In the west, it has been fashionable of late to argue that livestock are a drain on resources – and to suggest that they exacerbate global warming by producing methane gas as a by-product of rumination. In practice, in the modern world, both these criticisms are to some extent justified – but only because we do things badly; specifically, in general, because we have largely replaced traditional farming with industrial agribusiness.

A perfect system can include animals!

Thus conceptually, and traditionally, livestock falls into two categories. One the one hand there are the specialist herbivores, both ruminants (cattle and sheep the main ones, with goats and deer as minor players) and hind-gut digesters such as horses and rabbits (which are important in some economies). Camels may be seen as 'pseudo-ruminants', roughly similar to cattle in the way they deal with herbage, but differing in some details. In a state of nature the specialist herbivores derive most of their energy from cellulose – and this is an extraordinary trick to pull, and an extraordinarily valuable one, for cellulose is the most common organic polymer in nature. It is present and usually the prime component of every plant cell wall and so is common to all plants, growing in all circumstances. In nature at large there are megatonnes of it.

Human beings cannot derive significant quantities of energy from cellulose but by raising specialist herbivores they gain vicarious access to the cornucopia cellulose has to offer. Not every specialist herbivore can make use of every kind of leaf or stem – many plants are frankly toxic and/or protect themselves with thorns and whiskers and mucilage. But between them the domestic animals can derive nourishment from most common plants, which means that they can at least survive in most environments, and positively thrive in warm and rainy reasons when the plants are flourishing. So people who keep the right kinds of herbivores in the right proportions can survive in the most hostile landscapes. So it was that Job and his kin, in the Old Testament, lived very well in the desert (at least when he wasn't being assaulted by various plagues) with his mixed herd of cattle, sheep, goats, donkeys and camels – the proportions spelled out in the Book of Job and still to be found in some African herds.

Even when the plants are dying and the livestock are hungry and losing weight, they still can be killed for meat. The food they supply doesn't even need to be stored and carried. Obligingly, it moves itself around. When the plants are growing, or are about to, the animals give birth – and then their owners live on their milk. Indeed, in well-run desert communities, the animals are killed for meat *only* in times of drought, when the vegetation languishes and they are liable to die anyway. Hence the Jewish edict in Leviticus: not to eat meat and drink milk at the same meal. (I was once told off in a Kosher restaurant for drinking milky tea with a salt beef sandwich.) The Masai drink the blood of their cattle, bleeding them at judicious intervals. Nor are livestock kept only for food. They are the source of textiles (wool) and leather, of fertility, of fuel (cow dung) and their bones are used to make tools and furniture and even sometimes for building. In much of the world animals are the chief transport and pull tractors and harvesters for good measure. This is no anachronism. In some parts of the world and some economies animal power is still the best. The sacred cows of India produced calves that were castrated to make oxen for draught power – and these were almost free because the cows traditionally were not fed: they fed themselves from wayside weeds and crop residues. In much of Africa cattle and sometimes goats are the principal currency; and currency in relation to cattle, unlike currency in relation to banknotes, is real. Livestock, in short, may be a luxury in the middle class west but they are at the heart of some of the world's most venerable cultures.

The second category of livestock is the omnivores – pigs and poultry. Both can derive some energy from cellulose – though opinions differ on this, and the ability seems to vary somewhat from breed to breed. But on the whole, pigs and poultry eat the same kinds of things that humans eat. Yet their culinary standards are not high. So they are traditionally raised on leftovers – food wastes and crop surpluses. Hence they supplement the overall economy. Furthermore, pigs in particular are great cultivators, eating weeds and digging up the soil and fertilizing it. Indeed, they have often been kept for this alone, with their meat as a bonus. Chickens traditionally were moppers-up of wastes who offered a more or less continuous supply of eggs. Typically the hens themselves were eaten primarily when their laying days were over, in casseroles and pot roasts, while the cockerels made *coq au vin* – but only as an occasional treat.

Traditionally, then, the best land was used for growing crops – arable and horticulture – with pigs and poultry used to clear up the loose ends and for cultivation; and the fields beyond the cultivated area, including the uplands, which were often too steep or too wet or too dry to grow crops, were used for the herbivores; although, of course, sheep and cattle also came on to the arable fields in periods of fallow, to allow the ground to rest and to add fertility. Overall, such a system is highly efficient – not necessarily in cash terms, within the modern economy, but in biological terms, which matters far more. Indeed, as Kenneth Mellanby pointed out in his classic book of 1973, *Can Britain Feed Itself?*, if we reinstalled such a structure in Britain then we could certainly be self-reliant in food. Nowadays we can add refinements, too. Thus Martin Wolfe of the Organic Research Centre argues that all farming should be conceived as an exercise in agroforestry and at Wakelyns Farm in Suffolk he has shown a very neat way of doing this: growing a variety of valuable trees with benefit to other crops and potentially to livestock (although at present he has no livestock because of the shortage of labour). Aquaculture also has a big part to play (and fish are livestock too). Indeed in 2008 I helped to organize a conference in Oxford on national self-reliance and concluded, as others did, that with an updated, Mellanby-style structure and with some modern varieties we could feed ourselves easily (Tudge, 2009).

The methane problem also disappears if we farm in this traditional, commonsensical way. Cattle do indeed exhale methane (it comes out the front end, not the back end) but if they are fed only on grass, then of course they can exhale only the carbon that the grass itself has acquired from the atmosphere by photosynthesis. So the excretion of methane could be viewed merely as a form of recycling. There is no net increase in atmospheric carbon. Furthermore, if the grazing is controlled, then much of the carbon dioxide that the grass fixes by photosynthesis finishes up in the roots rather than the leaves and remains uneaten, and as the roots die so the carbon content of the soil is increased, so that a well-grazed field can be a net carbon *sink* – the very thing the world needs. Thus, grazing livestock (in a controlled way) should be win–win. Livestock becomes a menace only when, as now, we feed half the world's cereal and most of the world's soya to animals that should be getting the bulk of their nutrient from grass and leftovers. We burn vast quantities of oil to grow that cereal and fell forest to grow the soya – and so contribute twice over, and prodigiously, to global warming. Yet the cereal and soya we feed to the animals need not be grown at all – or if it is, then we could be eating it ourselves. Again we find it is not the animals that are at fault, but the system.

We can extend Kenneth Mellanby's principle. If all countries reinstalled the traditional structures then almost all of them could be self-reliant – and without wrecking their own environments or the climate. The world as a whole already produces more or less enough food to feed us all – distribution and justice are the problems, not productivity; and just by making life easier for farmers, and particularly for developing world farmers, we could easily add the 50 per cent needed to feed the 9.5 billion who will be with us by 2050. There is a huge serendipity here – because the same UN demographers who predict the rise to 9.5 billion also predict that numbers will stabilize at that figure, because also although *absolute* numbers continue to rise the *percentage* rise is falling off and by 2050 will be zero. Zero percentage rise means numbers will be steady. The population is then predicted to stay at around 9.5 billion for several hundred years and should then start going down – not because of wars and famines and other disasters but because that is the way demographic curves work. People are *choosing* to have fewer children and contraception makes this easier, such that in time the percentage rise will fall below replacement.

In short, the problem of feeding the world can now be seen to be finite (for the first time in 10,000 years) – and manageable. We need only to design agriculture specifically to feed people without wrecking the rest of the world – what I have called 'enlightened agriculture' (Tudge, 2004, 2007). Furthermore, enlightened agriculture brings with it two more huge serendipities. It produces plenty of plants, not much meat and maximum variety: and these nine words – 'plenty of plants, not much meat and maximum variety' – summarize all the most worthwhile nutritional theory of the past three decades. As the final bonus, these nine words also summarize the basic structure of all the world's greatest cuisines – Provence, Italy, Turkey, Persia, India, China. All of them are plant-based, using meat and fish only sparingly – for flavour and texture, as a garnish, for stock, and eaten in bulk only on the occasional feast day. These cuisines are unsurpassed – the basis of all worthwhile *haute cuisine*. But they are not the cuisines of the elite. They are peasant cooking, devised by ordinary people from whatever grew around – which basically was what farmers found easiest to grow. It all works beautifully. To eat well and sustainably, we don't even have to be austere. We just have to grow what grows best and take cooking seriously. In truth, the future belongs to the gourmet.

So we should be heaving a great sigh of relief. Instead we are panicking, resorting to ever more exotic technologies – and are told from on high that we need to *double* food output in the next 30 years or we will all starve, or at least a lot more people will. As things are, we are practising agriculture that seems designed to do the precise opposite of what is required – and are planning simply to do more of the same. It is all the grossest nonsense – and at the heart of this grossness, as always, is livestock. Livestock (like all of science), should be one of the great assets of humankind; but again (like science) it has become one of our principal threats.

What are we doing wrong?

Just about everything, is the short answer. 'Modern' industrialized farming of the kind that is anomalously called 'conventional' has rejected diversity in favour of monoculture. It does not recycle all inputs as traditional systems did and nature does. It relies on non-renewables – notably oil – as if there is no tomorrow, which indeed is becoming the case, and the manure that once was a prime reason for keeping livestock is routinely dumped (Kennedy, 2005).

In 'modern' systems, livestock is not kept to balance and supplement the crops. Increasingly, animals have become the *raison d'être*. Ruminants are still fed on grass and browse of course but they are also fed increasingly on concentrates – and so it is that cattle, pigs and poultry now consume half the world's wheat (the principal staple), 80 per cent of the world's maize, virtually all the barley that is not used for brewing and distilling and well over 90 per cent of the world's soya, now grown more and more in Brazil at the expense of the rainforest and the Cerrado. Animals raised in traditional ways – on grazing, browse, surpluses and leftovers – add to our food supply; they augment and reinforce whatever the plants provide. Animals fed on staples that could be feeding us are not supplementing our diet. They are competing with us.

Meat output continues to increase, so that by 2050, on present trends, livestock will be consuming basic food to feed 4 billion people. The additional

4 billion is roughly equivalent to the world population in the early 1970s when the United Nations held its first world conference to discuss what it saw as the global food crisis. Thus the world's effective population will not be 9.5 billion. It will be 13.5 billion – which is why the powers that be are now telling us that we need to double total output. At the same time, of course, the more livestock we raise the greater the problems they bring – excess greenhouses gases, destruction of forest, consumption of water. In such large numbers, in intensive units, they also become stewpots of infections, including several with the capacity to cause pandemic among human beings (see Greger, Chapter 10 this volume). Goodness me.

In short, if we are seriously interested in our own survival, and the future of the rest of the world, then the agriculture we have now is the grossest nonsense, ill-conceived and getting worse. Before we move on and ask what is to be done we should pause at least briefly to ask why we now behave so destructively, and seem prepared to continue in the same vein.

Why have we chosen to farm so perversely? Why don't we do what is sensible?

The answer lies with the economy. No, I am not a loonie-leftie Marxist (but then neither was Marx). I am a perfectly good capitalist. I reckon that the devices of capitalism, properly construed, are the best equipped of any conceived so far by which to manage our affairs. But capitalism as first conceived by Adam Smith and put into practice by the founding fathers of the United States – John Adams, Thomas Jefferson, James Madison and the rest – was tempered by morality, and by a keen and humble perception of what is physically possible. Adam Smith was a moralist before he was an economist and stressed 'natural sympathy' – what a Buddhist might call 'compassion'. The American Declaration of Independence, surely the greatest of all political manifestos, stressed the need to balance personal liberty with justice for all – 'We take it to be self-evident … etc.'.

But the modern form of capitalism has shaken off those constraints. It has produced the neoliberal, global, allegedly 'free' market. Morality, insofar as it is recognized at all, is intended to emerge from the market itself; what people will pay for is what they want and what they want must be good. The market itself is driven by the logic of cash – the perceived imperative to make as much of it as possible. This description may sound too crude by half but it is, in truth, an astonishingly crude concept. All human enterprises are conceived only as generators of cash. Agriculture is dragged in like all the rest – it has become, as the chill phrase has it, 'a business like any other'. The modern economy is also intended ostensibly to be maximally competitive in a horribly crude version of 'Darwinism', interpreted as 'strong bash weak'. Because the market is global and all-pervasive, an Indian family with two cows is in more or less direct, head-to-head competition with Smithfield, with its million-head pig cities, bottomless funds and battalions of lawyers.

Cynics may suggest that the point of the market is simply to make fat cats fatter and since all the statistics show that the gap between rich and poor has widened prodigiously this past few decades, both within and between countries, such cynicism may well seem justified. Not all the market's advocates are evil, however. Many simply believe, sincerely, in cash; that it is simply essential; that the more of it people have, the better they feel – that national wealth and national well-being go hand in hand; and – perhaps the most damaging illusion of all – that with enough cash we can always buy enough science and technology to dig ourselves out of any hole we might dig ourselves into. In the interests of making a living, many scientists and technologists encourage politicians and financiers in this belief, even though they must know better.

The effects on agriculture are disastrous. The logic of the global cash market is precisely at odds with the requirements of Enlightened Agriculture – farming that is seriously intended to feed us. Again, livestock is at the heart of the matter.

Whatever they are producing, producers who seek to maximize profit, as has become imperative, must do three things. First, they must maximize turnover: 'Pile 'em high and sell 'em cheap', as Tesco's founder Jack Cohen put the matter. Second (as for Cohen) they must 'add value' – make whatever is cheap as dear as possible before selling it. Third, they must cut costs.

In agriculture, maximizing turnover means increasing the area farmed, and maximizing yield. The world's most fertile land has long been spoken for and now, to make space, we are cutting down what is left of the rainforest. We are seeking to maximize yield – of everything – by higher and higher inputs and through breeding programmes that are more and more narrowly focused on yield and, for the most part, steadily reduce the underlying diversity. But if we assume that the world needs only to raise total output by about 50 per cent, then the obsession with yields seems somewhat misguided. If we acknowledge that farmers – and particularly developing world farmers – could increase yield substantially with their present techniques if only the prices were guaranteed (as in pre-global market days was often the case) so that they knew they could justify the cost of inputs, then there is absolutely no case for the more and more fancy technologies that are designed to squeeze the last gram of starch out of wheat and the last drop of milk out of cows. In particular, the principal case for genetic engineering collapses absolutely, particularly when applied to livestock – although this has become the talisman, the jewel in the crown of modern agricultural endeavour. If we take global warming and the world's water shortage seriously we see that it is far more important to devise crops, live-stock and above all *systems* that are resilient to change. Mere output is far less of a priority. But the conservative, median yields provided by systems that are designed to be flexible are not maximally profitable. Long-term resilience does not lead to short-term profit. So that is not the emphasis.

Value-adding includes excess packaging and processing and French beans out of season whisked halfway around the world to be sold in English suburbs – and all this is wasteful but it is not the main problem. The main issue here by far – again – is livestock. For modern agriculture focuses, more and more, on increasing the output of meat. This is not good nutri-tionally – a small amount of meat is highly desirable but too much is liable to be harmful – and neither is it necessary gastronomically, as the great cuisines demonstrate. The real reason for maximizing livestock is to add value. It would be far *too* easy to feed everybody well if everyone appreci-ated the traditional, plant-based cuisines. But this would not be maximally profitable. If everyone has enough then a market is said to be 'saturated' – whether the commodity in question is wheat or houses or diamonds – and for the suppliers, a saturated market is not good news at all. It's shortage that keeps the prices up; and besides, in the modern ultra-competitive economy, the imperative for every business is to *grow* – which means producing more and more. Livestock provides the perfect antidote. The market for meat can *never* be saturated. If people have enough meat, then simply throw most of the carcass away – put it into cat food or other animal feed – and sell only the steaks and chops. These days there's the additional back-up of biofuel.

Of course, the industrial farmers and the corporates that supply this meat, and the governments that support them in this, argue that they are merely meeting 'demand' and hence behaving democratically. This argument is supported by more spurious biology – the notion that people are basically carnivores, and eat more and more meat as soon as they are rich enough to do so. In truth, meat has been sold, sold, and sold again this past half century, as energetically as any other lucrative commodity, from motor cars to soap powder. To be sure, people who become affluent after years of deprivation do tend to eat more meat – we are good opportunist omnivores, after all – and for a time, meat tends to become associated with social status. But people who really know food do not focus on meat – as the great cuisines demon-strate; and people who are too obviously affluent to bother to prove it do not necessarily eat much meat. The mandarins of old China were rich beyond the dreams of Croesus – and they gave the flesh of ducks to the servants and ate only the skin, in the form of Peking duck. They knew that meat is just for flavour – and the skin is the most flavoursome. Among the modern, western, affluent middle class, vegetarianism is positively chic. The way to flaunt status these days is to demonstrate that you have time and energy to devote to cooking. Meat eating is not an evolved and ineradicable obsession after all. For the most part, it is largely fashion. But like all fashions, for those in the driving seat, it is immensely lucrative; and when profit is God, the fashion must be kept going even when it threatens the whole world.

But the final imperative of profit – to cut costs – is even more pernicious. It results in cut-price husbandry. In the case of livestock, this is extremely dangerous. All the foulest epidemics of Britain of the past 30 years – bovine spongiform encephalopathy (BSE), the worst foot and mouth outbreak ever, then swine fever, then more foot and mouth and the recent brush with bird flu and the current threat of swine flu – have sprung from the perceived imperative to keep animals in vast numbers, as cheaply as possible.

Worst of all, though, is that to reduce costs, labour must be cut – for labour, in traditional systems, is generally the most expensive input (even slaves must be fed). This affects farming, and the world as a whole, in two ways. First, the polyculture that is the basis of Enlightened Agriculture is intricate. This means that perforce, it is labour-intensive. If labour is cut, then polyculture must give way to monoculture. Monoculture is not especially 'efficient' as often claimed – not if you cost all the inputs, and the collateral damage. But it is simple and so can be done by machines and industrial chemistry, aided by biotech, rather than by human ingenuity. In industrialized systems, too, there are tremendous advantages of scale – and so the machines get bigger and bigger and so do the farms, and the agrochemical and seed companies become more and more consolidated. Complexity, and hence diversity, flexibility and long-term resilience go out of the window. But it's cheaper that way – and that is all that is deemed to count.

Even more importantly, despite the best efforts of governments and industry to cut its labour force, agriculture is still the world's biggest employer. In the developing world as a whole 60 per cent of the working population live on the land. India alone has more than 250 million farmers, and 600 million people rely directly on those farms – far more than the total population of the newly expanded European Union (EU). Governments and 'experts' who advocate the wholesale industrialization of developing world agriculture insist that there are 'alternative industries' but in truth there are not, and in reality there can never be. The world does not have enough resources to create industries that could employ so many people, and certainly not usefully. Many people in slums these days earn a living of a kind by recycling old tyres and polythene bags but although that is useful up to a point it can hardly be preferred to farming. In short, the urge to cut costs and then cut them again, by whatever means, is creating unemployment on a scale that matches all the other disasters of the world. What price the much-vaunted 'war on poverty'? The industrialization of agriculture that the west calls progress may be poverty's greatest proximal cause.

Again, livestock is at the heart of all the trouble. Small-scale livestock farming, often practiced most ingeniously, is vital. Cattle in the developing world are multi-purpose – calves, milk, dung for fuel and fertility, transport and draught; but, traditionally, they are fed only on crop residues, or are taken out on leads to feed from the roadside verges. This may not be ideal for all kinds of reasons but in principle such thrift is precisely what is needed – and the task of science, when it has a role at all, is to build on the traditional systems and make them work; not to sweep them aside in pursuit of the wild and unrealizable dream that we have come to equate with 'progress'. We need to recognize again the absolute need for a sound rural economy that continues to employ vast numbers of people – not necessarily the 60 per cent of the current developing world, but certainly at least a fifth. We need, in short, what a truly progressive farming friend of mine calls 'the Agrarian Renaissance'. We should be using our science and technology not to make fat cats fatter but to make small-scale farming easier and agrarian living more agreeable. Much of the necessary technology already exists. Much of it is low tech but high tech is highly relevant too – from rotavators that take the grunt out of small-scale cultivation (and could be powered by solar-generated electricity) to IT, which ensures that no one, wherever they work, need be cut off from the rest of the world (unless they choose to be).

So what's to be done?

It is often argued that the free market economy could still solve our problems, given time, since it depends upon consumers: and consumers can change market practice by their spending patterns. Many consumers have objected to various forms of intensive livestock production and in response, some supermarkets are now seeking to improve animal welfare – buying only from farmers with high welfare standards. This is indeed encouraging but there is cause to wonder how far this can be taken, and how quickly. Would it really be possible to reform the corporates who now control the world's food supply and the governments who support those corporates (because they are presumed to promote 'economic growth'), just by exerting such consumer pressure? Or do we simply need a quite different way of organizing the world?

Many now believe, and some always have believed, that we – humanity – cannot leave our future and our children's future to governments, as we are inclined and conditioned to do. It is a mistake to suppose that a government must be on our side just because it has been elected, or that it knows what it is doing just because it has recourse to experts. Governments choose their own experts. If we really care about the future, and our children's future, and other people and their children, and other species, then we have to take

matters into our own hands. We need a 'people's movement'. The po...
not to mount a revolution, which would hurt people and would be mo...
or less bound to fail, but to effect a Renaissance: to build something better
from what is around, and allow the status quo to whither on the vine. For my
part, together with a few well-informed friends, I helped in 2008 to found
an outfit called 'LandShare' whose prime aim is to help get people back on
the land. More specifically, in 2009 (as a project of LandShare) I launched
'The Campaign for Real Farming', intended to bring farmers, retailers, and
indeed everyone who gives a damn on board, truly to create an alternative
food supply chain, which will be run as Abraham Lincoln put the matter 'by
the people and for the people'. As I said, this is not a loonie-leftie project. It
partakes entirely of the spirit and many of the basic ideas that inspired the
foundation of the United States – which, after all, on a significant point of
detail, Jefferson envisaged as 'a nation of small farmers'.

Maybe the Campaign for Real Farming is a forlorn gesture – but not
as forlorn as attempts to improve Europe's Common Agricultural Policy,
which I have been invited to take part in recently; or the wranglings of the
G20. Such tweaking and nibbling gets nowhere near the roots of the problems, and is not designed to. It begins with the assumption that the basis of
the present-day economy and power structure is OK, or at least is a given –
that it is the natural way of things for nation states to jockey for position in
a global market dominated by corporates and banks. But this is just not so.
If we are to do anything worthwhile to help ourselves and the world we first
have to recognize that *Homo sapiens* is a biological species and the Earth is our
habitat, and our only habitat, now and forever. The 'real' reality in the world is
not the economy and the power struggles of nations and political parties but
the laws of physics and the rules of biology – that, and the fact that despite
everything, human beings are basically intelligent and nice. If we don't recognize all this, and act accordingly, then we really missed our chance.

For more on the Campaign for Real Farming, please log in to: http://
campaignforrealfarming.blogspot.com.

References

Kennedy, Jr, R. F. (2005) *Crimes Against Nature*, Harper Collins, New York
Mellanby, K. (1973) *Can Britain Feed Itself?* Merlin Press, London
Tudge, C. (2004) *So Shall We Reap*, Penguin Books, London
Tudge, C. (2007) *Feeding People is Easy*, Pari Publishing, Pari, Italy
Tudge, C. (2009) *Can Britain Feed Itself? Should Britain Feed Itself?* LandShare, Oxford

2

Water Footprint of Animal Products

Arjen Y. Hoekstra

Introduction

One single component in the total water footprint of humanity stands out: the water footprint related to the consumption of animal products. About 85 per cent of humanity's water footprint is related to the consumption of agricultural products; 10 per cent relates to industrial products and only 5 per cent to domestic water consumption (Hoekstra and Chapagain, 2007, 2008). Within the category of agricultural products, animal products generally have a much larger water footprint per kilogram or calorific value than crop products. This means that if people consider reducing their water footprint, they are advised to look critically at their diet rather than at their water use in the kitchen, bathroom and garden. Wasting water never makes sense, so saving water at home when possible is certainly advisable, but if we limit our actions to water reductions at home, many of the most severe water problems in the world will hardly be lessened. The water in the Murray-Darling basin in Australia is so scarce mostly because of water use for the production of various types of fruits, vegetables, cereals and cotton. The Ogallala Aquifer in the American Midwest is gradually being depleted because of water abstractions for the irrigation of crops like maize and wheat. Many of the grains cultivated in the world are not for human consumption but for animals. In the United States, for example, 68 per cent of the grains consumed are used for animal feed (Millstone and Lang, 2003). Animal products have a relatively large water footprint because of the water needed to grow their feed rather than the water volumes required for drinking. From a water saving point of view it is obviously more efficient to eat the crops directly rather than indirectly by having them first processed into meat. Surprisingly, however, little attention is paid among scientists and policy makers to the relationship between meat consumption and water use.

Consumers can reduce their direct water footprint – in other words, their home water use – by installing water-saving toilets, applying a water-saving showerhead, etc. For reducing their indirect water footprint – that is the water consumption behind the production of food and other consumer products – they have two options. One option is to substitute a consumer product that has a large water footprint with a different type of product that has a smaller water footprint. Eating less meat or becoming vegetarian is one example, but one can also think of drinking tea instead of coffee, or, even better, plain water. Wearing artificial fibre rather than cotton clothes also saves a lot of water. But this approach of substitution has limitations, because many people do not easily shift from meat to being vegetarian and people like their coffee and cotton. A second option is that people stick to the same consumption pattern but select the beef, coffee or cotton that has a relatively low water footprint or that has its footprint in an area that does not have high water scarcity. This means, however, that consumers need proper information to make that choice. Since this sort of information is generally not available, this in turn asks for an effort from businesses to create product transparency and an effort from governments to install the necessary regulations. Currently we are far removed from a situation in which we have relevant information about the environmental impact of one piece of beef compared to another piece. The water footprint of beef, however, greatly varies across production systems and countries and strongly depends on feed composition. The same holds for other animal products.

In this chapter, I will start by comparing the water footprints of a number of animal products with the water footprints of crops. Second, I will compare the water footprint of a meat-based diet with a vegetarian diet. Then I will show that understanding the relation between food consumption and the use of freshwater resources is no longer a local issue. Water has become a global resource, whereby – due to international trade – food consumption in one place often affects the water demand in another place. Finally, I will argue for product transparency in the food sector, which would allow us to better link individual food products to associated water impacts, which in turn can drive efforts to reduce those impacts.

The water footprint of animal products

The water footprint concept is an indicator of water use in relation to consumer goods (Hoekstra, 2003). The concept is an analogue to the ecological and the carbon footprint, but indicates water use instead of land or fossil energy use. The water footprint of a product is the volume of fresh water used to produce the product, measured over the various steps of the production chain. Water use is measured in terms of water volumes consumed (evaporated) or polluted. The water footprint is a geographically explicit indicator that not only shows volumes of water use and pollution, but also the locations. A water footprint generally breaks down into three components: the blue, green and grey water footprint. The blue water footprint is the volume of fresh water that is evaporated from the global blue water resources (surface and groundwater). The green water footprint is the volume of water evaporated from the global green water resources (rainwater stored in the soil). The grey water footprint is the volume of polluted water, which is quantified as the volume of water that is required to dilute pollutants to such an extent that the quality of the ambient water remains above agreed water quality standards (Hoekstra and Chapagain, 2008).

In order to understand better the water footprint of an animal product, we had better start by explaining the water footprint of feed crops. The water footprint (m³/ton) of a crop when harvested from the field is equal to the total evapotranspiration from the crop field during the growing period (m³/ha) divided by the crop yield (ton/ha). The crop water use depends on the crop water requirement on the one hand and the actual soil water available on the other hand. Soil water is replenished either naturally through rainwater or artificially through irrigation water. The crop water requirement is the total water needed for evapotranspiration under ideal growth conditions, measured from planting to harvest. It obviously depends on the type of crop and climate. Actual water use by the crop is equal to the crop water requirement if rainwater is sufficient or if shortages are supplemented through irrigation. In the case of rainwater deficiency and the absence of irrigation, actual crop water use is equal to effective rainfall. The green water footprint refers to the part of the crop water requirement met through rainfall; the blue water footprint is the part of the crop water requirement met through irrigation. The grey water footprint of a crop is calculated as the load of pollutants (fertilizers, pesticides) that are leached from the field to the groundwater (kg/ha) divided by the ambient water quality for the chemical considered (g/l) and the crop yield (ton/ha).

The water footprint of an animal at the end of its lifetime can be calculated based on the water footprint of all feed consumed during its lifetime and the volumes of water consumed for drinking and, for example, cleaning the sheds. One will have to know the age of the animal when slaughtered and the diet of the animal during its various stages of life. The water footprint of the animal as a whole is allocated to the different products that are derived from the animal. This allocation is done on the basis of the relative value of the various animal products, avoids double counting and assigns the largest shares of the total water input to the high-value products and smaller shares to the low-value products.

Table 2.1 shows the water footprint for a number of common animal products and for a number of food crops as well. The numbers in the table

Table 2.1 *The global-average water footprint of animal products versus crops*

Animal product	litre/kg	Crop	litre/kg
Bovine leather	16,600	Rice	3400
Beef	15,500	Groundnuts (in shell)	3100
Sheep meat	6100	Wheat	1300
Cheese	5000	Maize	900
Pork	4800	Apple or pear	700
Milk powder	4600	Orange	460
Goat meat	4000	Potato	250
Chicken	3900	Cabbage	200
Eggs	3300	Tomato	180
Milk	1000	Lettuce	130

Source: Hoekstra and Chapagain (2008)

are global averages. These averages show that, as expected, animal products are more water-intensive than food crops. However, what the averages hide is that there is a very large variation of the water footprint for each of the products shown in the table. Knowing these differences is essential if one has already chosen to buy a certain product but not yet chosen which of the various options still remain. One piece of beef is simply not equal to the other one, even though the taste and all other measurable characteristics are the same. The history may be different.

Consider the water footprint of beef. Let us consider an industrial beef production system where it takes three years before the animal is slaughtered to produce 200kg of boneless beef. Suppose that the animal has consumed 1300kg of grains (wheat, oats, barley, corn, dry peas, soybean meal and other small grains), 7200kg of roughages (pasture, dry hay, silage and other roughages), 24 cubic metres of water for drinking and 7 cubic metres of water for cleaning sheds etc. This means that to produce 1kg of boneless beef, we have used about 6.5kg of grain, 36kg of roughages, and 155 litres of water (only for drinking and servicing). Producing the volume of feed has cost about 15,300 litres of water. The water footprint of 1kg of beef thus adds up in this case to 15,500 litres of water. This still excludes the volume of polluted water that may result from leaching of fertilizers in the feed crop field or from surplus manure reaching the water system. The example given here can be considered as more or less a global average case. The water footprint of beef will vary strongly, however, depending on the production region, feed composition and origin of the feed ingredients. In a grazing system, cows will eat more grass and roughage and less grains. When slaughtered after three years they will have gained less weight and provide less meat. In terms of total volume, the water footprint of beef from an industrial system will thus be lower than for a grazing system, but maybe more important is that the water footprints are completely different in terms of where the water is sourced. The water footprint of beef from an industrial system may partly refer to irrigation water (blue water) to grow feed in an area remote from where the cow is raised. This can be an area where water is abundantly available, but it may also be an area where water is scarce and where minimum environmental flow requirements are not met due to overdraft. The water footprint of beef from a grazing system will mostly refer to green water used in nearby pastures. If the pastures used are either dry- or wetlands that cannot be used for crop cultivation, the green water flow turned into meat could not have been used to produce food crops instead. If, however, the pastures can be substituted by cropland, the green water allocated to meat production is no longer available for food-crop production. This explains why the water footprint is to be seen as a multidimensional indicator. One should not only look at the total water footprint as a volumetric value, but one should also consider the green, blue and grey components separately and look at where each of the water footprint components are located. The social and ecological impacts of water use at a certain location depend on the scarcity and alternative uses of water at that location.

Water consumption in relation to diet

Since food consumption gives the most important contribution to the water footprints of people, even in industrialized countries, dietary habits greatly influence the associated water footprint. In industrialized countries the average calorie consumption today is 3400kcal per day; roughly 30 per cent of that comes from animal products. When we assume that the average daily portion of animal products is a reasonable mix of beef, pork, poultry, fish, eggs and dairy products, we can estimate that 1kcal of animal product requires on average roughly 2.5 litres of water. Products of vegetable origin, on the other hand, require roughly 0.5 litres of water per kcal, this time assuming a reasonable mix of cereals, pulses, roots, fruit and vegetables. Under these circumstances, producing the food for one day costs 3600 litres of water (see Table 2.2). In developing countries the average consumption is lower: about 2700kcal per day per person, only 13 per cent of which is of animal origin. Such a diet costs 2050 litres of water per day. These numbers are averages over averages, because, firstly,

Table 2.2 *The water footprint of two different diets for industrialized and developing countries*

	Meat diet	kcal/ day[a]	litre/ kcal[c]	litre/ day	Vegetarian diet	kcal/ day[b]	litre/ kcal[c]	litre/ day
Industrialized countries	Animal origin	950	2.5	2375	Animal origin	300	2.5	750
	Vegetable origin	2450	0.5	1225	Vegetable origin	3100	0.5	1550
	Total	3400		3600	Total	3400		2300
Developing countries	Animal origin	350	2.5	875	Animal origin	200	2.5	500
	Vegetable origin	2350	0.5	1175	Vegetable origin	2500	0.5	1250
	Total	2700		2050	Total	2700		1750

Notes:
[a] The numbers are equal to the actual daily caloric intake of people in the period 1997–1999 (FAO, 2003).
[b] This example assumes that the vegetarian diet still contains dairy products.
[c] For each food category, a rough estimate has been made by taking the weighted average of the water footprints (litre/kg) of the various products in the food category (from Hoekstra and Chapagain, 2008) divided by their respective caloric values (kcal/kg); the estimate for food from vegetable origin coincides with the estimate made by Falkenmark and Rockström (2004); for food from animal origin, the latter use a higher value of 4 litres/kcal.

total caloric intakes and meat fractions assumed vary between and within nations and, second, the water requirements actually vary across production regions and production systems. The averages shown here mainly function to make a comparison between the water footprints of a meat-based versus a vegetarian diet. For the vegetarian diet we assume that a smaller fraction is of animal origin (not zero, because of dairy products still consumed), but keep all other factors equal. In the example for the industrialized countries this reduces the food-related water footprint by 36 per cent (see Table 2.2). In the case of developing countries the switch to vegetarian diet saves 15 per cent of water. Keeping in mind that for the 'meat eater' we had taken the average diet of a whole population and that meat consumption varies within a population, larger water savings can be achieved by individuals that eat more meat than the average person.

From the above figures it is obvious that consumers can reduce their water footprint by reducing the volume of their meat consumption. Alternatively, however, or in addition, consumers can reduce their water footprint by being more selective in the choice of which piece of meat they pick.

Chickens are less water-intensive than cows, and beef from one production system cannot be compared in terms of associated water impacts to beef from another production system.

The relation between food consumption and water use: A global issue

Protecting freshwater resources can no longer be regarded as an issue for individual countries or river basins. Let us take Europe as an example. The water footprint of Europe – the total volume of water used for producing all commodities consumed by European citizens – has been significantly externalized to other parts of the world. Europe is a large importer of crops like sugar and cotton, two of the thirstiest crops. Europe also imports large volumes of feed, like soybean from Brazil. European consumption relies heavily on water resources available outside Europe. How is Europe going to secure its future water supply? China and India are still largely water self-sufficient, but with rising food demand and growing water scarcity within these two major developing countries, one will have to expect a larger demand for food imports and thus external water demand.

Although in many countries most of the food still originates from the country itself, substantial volumes of food and feed are internationally traded. As a result, all countries import and export water in virtual form, in other words, in the form of agricultural commodities. Within Europe, France is the only country with a net export of virtual water. All other European countries have net virtual water import: they use some water for making export products but more water is used elsewhere to produce the commodities that are imported. Europe as a whole is a net importer of virtual water. Europe's water security thus strongly depends on external water resources. Related to this, a substantial proportion of existing problems of water depletion and pollution in the world relates to export to Europe.

Countries like Australia, Canada, the United States, Brazil and Argentina have a net export of water in virtual form because of trade in animal products. Other countries, like Japan, China, Italy and Russia have a net import (see Table 3.3). Total international virtual water flows related to global trade in animal products add up to 275 billion m^3/yr, a volume equivalent to about half the annual Mississippi run-off.

Not only are livestock and livestock products internationally traded, but also feed crops. In trade statistics, however, it is difficult to distinguish between food and feed crops, because they are mostly the same crops but the application is different. Worldwide, trade in crops and crop products results in international virtual water flows that add up to 979 billion m^3/yr

Table 2.3 *Import and export of virtual water related to the trade in animal products*

	Import	Export	Net import
	million m³/yr		
Argentina	811	4180	−3370
Australia	745	26,400	−25,600
Bangladesh	86	652	−566
Brazil	1910	11,900	−10,000
Canada	4950	17,400	−12,500
China	15,200	5640	9610
Egypt	1470	221	1250
France	11,800	13,200	−1390
Germany	16,100	17,400	−1370
India	343	3,410	−3060
Indonesia	1670	371	1300
Italy	28,300	14,900	13,400
Japan	20,300	955	19,400
Jordan	462	165	297
Korea Rep.	6,100	3930	2170
Mexico	13,400	5760	7660
Netherlands	7850	15,100	−7300
Pakistan	98	612	−514
Russia	12,200	2500	9740
South Africa	1020	1310	−293
Spain	5970	8540	−2570
Thailand	1760	2860	−1100
United Kingdom	10,200	3790	6380
USA	32,900	35,500	−2560

Source: Hoekstra and Chapagain (2008)

(Hoekstra and Chapagain, 2008), which is equivalent to 1.5 to 2 times the annual run-off of the Mississippi.

Even now water is still mostly considered as a local or regional resource, to be managed preferably at catchment or river basin level. However, this approach obscures the fact that many water problems are related to remote consumption elsewhere. Water problems are an intrinsic part of the world's economic structure in which water scarcity is not translated into costs to either producers or consumers; as a result there are many places where water resources are depleted or polluted, with producers and consumers along the supply chain benefiting at the cost of local communities and ecosystems. It is unlikely that consumption and trade are sustainable if they are accompanied by water depletion or pollution somewhere along the supply chain. Typical products that can often be associated with remote water depletion and pollution are cotton and sugar products. For animal products it is much more difficult to tell whether they relate to such problems, because animals are often fed with a variety of feed ingredients and feed supply chains are difficult to trace. So unless we have milk, cheese, eggs or meat from an animal that was raised locally and that grazed locally or was otherwise fed with locally grown feedstuffs, it is hard to say which product has affected the world's scarce freshwater resources. The increasing complexity of our food system in general and the animal product system in particular hides the existing links between the food we buy and the resource use and associated impacts that underlie it.

Product transparency in the food sector

In order to know what we eat, we need a form of product transparency that is currently completely lacking. It is reasonable that consumers (or consumer organizations on their behalf) have access to information about the history of a product. Since in this chapter we focus on the relation between food and water resources use and impacts, the most relevant question is: how water-intensive is a particular product that is for sale and to what extent does it relate to water depletion and/or pollution? Establishing a mechanism that makes sure that such information is available is not an easy task. It requires a form of accounting along production and supply chains that accumulates relevant information all the way to the end point of a chain.

Figure 2.1 shows the various steps in the supply chain of an animal product. The chain starts with feed crop cultivation and ends with the consumer. In each step of the chain there is a direct water footprint, which refers to the water consumption in that step, but also an indirect water footprint, which refers to the water consumption in the previous steps. By far the biggest contribution to the total water footprint of all animal products comes from the first step: growing the feed. This step is the furthest removed from the consumer, which explains why consumers generally have little notion that animal products require a lot of water. Besides, the feed will

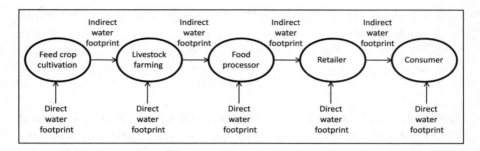

Figure 2.1 The direct and indirect water footprint in each stage of the supply chain of an animal product

often be grown in areas completely different from where the consumption of the final product takes place.

Governments that stress in particular their interest in 'sustainable consumption' may translate this interest into their trade policy. The Dutch government, for example, given the fact that about 80 per cent of the total Dutch water footprint lies outside its own territory, may strive towards more transparency about the water impacts of imported products. Achieving such a goal will obviously be much easier if there is international cooperation in this field. In cases where industrialized countries import feed from developing countries, the former can support the latter within the context of development cooperation policy in reducing the impacts on local water systems by helping to set up better systems of water governance.

Business can have a key role as well, particularly the large food processors and retailers. Since they form an intermediary between farmers and consumers they are the ones that have to pass on key information about the products that they are trading. As big customers they can also put pressure on farmers to actually reduce their water footprint and require them to provide proper environmental accounts. If it comes to water accounting, there are currently several parallel processes going on in the business world. First of all, there is an increasing interest in the water use in supply chains, on top of the traditional interest in their own operational water use. Second, several companies, including for instance Unilever and Nestlé, have started to explore how water footprint accounting can be practically implemented. Some businesses think about extending their annual environmental report with a paragraph on the water footprint of their business. Others speak about water labelling of products (either on the product itself or through information available online) and yet others explore the idea of water certification for companies. The interest in water footprint accounting comes from various business sectors, ranging from the food and beverage industry to the apparel and paper industries, but within the food industry there is still little interest the most water-intensive form of food: animal products.

Conclusion

Remarkably little is known about the relation between animal products and water resources protection. There is no national water plan in the world that even addresses the issue that meat and dairy products are among the most water-intensive consumer products, let alone that water policies somehow involve consumers or the food industry in this respect. Water policies are often focused on 'sustainable production' but they seldom address 'sustainable consumption'. The advantage of involving the whole supply chain is that enormous leverage can be created to establish change. The latest World Water Development Report by the United Nations (World Water Assessment Programme, 2009) is nearly 350 pages long and includes the word 'meat' just 15 times. The analysis does not go deeper than stating that global meat demand will increase and thus water demand as well. The same bias can be seen in scientific literature. One can find an endless number of publications on 'water and agriculture' and 'water for food' but they generally address the issue of water-use efficiency within agriculture (more crop per drop), hardly ever the issue of water-use efficiency in the food system as a whole (more kcal per drop).

Wise water governance is a shared responsibility of consumers, governments and businesses. As follows from the previous sections, each of those players has a different role:

- consumers (or consumer and environmental organizations) may demand more product transparency of animal products by business and governments, so that they are better informed about associated water resources use and impacts;
- consumers can choose to consume fewer animal products and – when still consuming those products – they can choose, whenever proper information allows, from the meat, eggs and dairy products that have a relatively low water footprint or for which this water footprint has no negative environmental impacts;
- national governments can – preferably in the context of an international agreement – put regulations in place that urge businesses along the supply chain of animal products to cooperate in creating product transparency;
- national governments can tune their trade and development cooperation policies towards their wish to promote consumption of and trade in sustainable products;

- businesses, particularly big food processors and retailers, can use their power in the supply chain to effect product transparency of animal products;
- businesses can cooperate in water labelling, certification and benchmarking schemes and produce annual water accounts that include a report of the supply-chain water footprints and associated impacts of their products.

References

Falkenmark, M. and Rockström, J. (2004) *Balancing Water for Humans and Nature: The New Approach in Ecohydrology*, Earthscan, London

FAO (2003) FAOSTAT, Food and Agriculture Organization, available at http://faostat. fao.org, last accessed 23 December 2009

Hoekstra, A. Y. (ed.) (2003) *Virtual Water Trade: Proceedings of the International Expert Meeting on Virtual Water Trade*, UNESCO-IHE Value of Water Research Report Series No 12, Delft, The Netherlands

Hoekstra, A. Y. and Chapagain, A. K. (2007) 'Water footprints of nations: Water use by people as a function of their consumption pattern', *Water Resources Management*, vol 21, issue 1, pp35–48

Hoekstra, A. Y. and Chapagain, A. K. (2008) *Globalization of Water: Sharing the Planet's Freshwater Resources*, Blackwell Publishing, Oxford

Millstone, E. and Lang, T. (2003) *The Atlas of Food: Who Eats What, Where and Why*, Earthscan, London

World Water Assessment Programme (2009) *The United Nations World Water Development Report 3: Water in a Changing World*, UNESCO Publishing, Paris/Earthscan, London

3

Livestock and Climate Change

Tara Garnett

The process of feeding ourselves has always had damaging consequences for our land (Diamond, 2005) and the animals we have reared. We have drained peatlands and chopped down forests to provide us with fertile agricultural soil, and we have used the animals in our care in often unspeakably cruel ways (Thomas, 1991).

Until now though, we have always been able to expand our way out of the problem. There was always new land to use. The situation today is very different: there are more of us – but there are no more planets to exploit. By 2050 there will be 9 billion or more of us on this earth and each one of us will need to eat. What we eat, and how we produce it, has a profound impact upon the long-term sustainability of the planet. Worryingly however, the patterns of production and consumption that we in the developed world have adopted, and that the developing world is rapidly taking up, have potentially catastrophic consequences.

One of the major causes for concern is the rapid growth in the production and consumption of foods of animal origin. Apart from the potential welfare implications of how we actually rear these animals, the growth in livestock farming has damaging consequences for biological diversity, for water extraction and use, for soil and air pollution – and for climate change, which is the primary focus of this chapter.

We show that tackling climate changing emissions will require us to reduce the number of animals we rear and to change the way we rear them; and we will need to eat fewer foods of animal origin. This said, livestock production can also very much form a part of the solution – farm animals can help create a resilient, sustainable, biodiverse food system, if we rear them in the right way and at moderate scales.

Finally, the livestock-climate issue cannot and must not be seen as standing in opposition to development objectives, and in particular to the right of all people to food security. We need to develop strategies that explicitly link the goals of ensuring food security, with that of achieving greenhouse gas (GHG) emission reductions from the agricultural and food sectors.

Climate change and international development: Twin problems

The climate is changing and most of the change is caused by human activity. The latest report (2007) by the Intergovernmental Panel on Climate Change (IPCC) concludes that 'Warming of the climate system is unequivocal...' what is more, 'Most of the observed increase in globally averaged temperatures since the mid-20th century is *very likely*[1] due to the observed increase in anthropogenic greenhouse gas concentrations.' (original emphasis)

In the last 100 years we have seen a global rise in temperature of 0.74°C. 11 of the last 12 years (1995–2006) have been the warmest since records began in 1850 (Schneider et al, 2007). Under 'business-as-usual' (BAU) scenarios, the IPCC warns that we are likely to see a temperature rise of about 3°C by 2100 relative to the end of the 20th century, within a possible range of 2°C to 4.5°C (IPCC, 2007). Other more recent studies suggest that the upper end of the estimate is more likely (Anderson and Bows, 2008).

It is generally accepted that a rise of 2°C above pre-industrial levels, equivalent to a concentration of CO_2 equivalent (CO_2e) in the atmosphere above 450 parts per million (ppm), delivers the probability of 'dangerous climate change' (Schellnhuber et al, 2006). We could then experience major irreversible system disruption, with hypothetical examples including a sudden change in the Asian monsoon or disintegration of the west Antarctic ice sheet (Schneider et al, 2007). Note that even at 450ppm there is only a 50 per cent chance of keeping the temperature rise to 2°C or lower (Hadley Centre, 2005). What is more, even if the world stopped emitting any more GHGs as of now, we would still be 'committed', due to time lags in the Earth's climate mechanisms, to a rise of 1°C by the end of the century (or 0.1°C per decade). In short, we have very little room left for manoeuvre.

Compounding the problem is poverty. Around 2.5 billion people worldwide have no access to proper sanitation, more than a third of the world's growing urban population live in slums and as many as 100 million people live in absolute poverty (United Nations, 2008). Over a billion people are malnourished, a figure that includes one in every four children in developing countries (FAO, 2009a). While we in the developed world can afford to make drastic changes in our lifestyles (even if we may not want to) to tackle climate change, there is an urgent need for people in the developing world to *raise* their standard of living – and this will mean increases in GHG emissions.

The implications for the developed world are clear. If we are to keep the global concentrations of greenhouses gases in the atmosphere to below 450 CO_2e ppm, *and* if the developing world economies are to achieve a reasonable standard of living, then rich countries, who contribute the bulk of present and historical greenhouse gas (GHG) emissions, need to reduce their emissions by 80 per cent or more (Committee on Climate Change, 2008), with the IPCC's Fourth Assessment report suggesting that a reduction of up to 95 per cent may even be needed (Gupta et al, 2007). Critically, we also need to be taking steps to reduce our emissions right now; the longer we put off taking action, the harder it will be to keep emissions beneath the 450ppm threshold (Stern, 2006).

Climate change and the role of food

The food chain contributes significantly to GHG emissions, both at national (UK) and international levels. Estimates vary, with ranges given from about 19 per cent (Garnett, 2008) of UK to 31 per cent (European Commission, 2006) of EU total emissions (reflecting, among other things, different methodological approaches), but clearly the impact is considerable.

Emissions arise at all stages in the food production cycle, from the farming process itself (and associated inputs) through manufacture, distribution and retailing, to the storing and cooking of food in the home or catering outlet. At the farming stage, the dominant GHGs are nitrous oxide (N_2O) from soil and livestock processes (faeces and urine) and methane (CH_4) from ruminant digestion and rice cultivation. Carbon dioxide (CO_2) emissions arise from the fossil fuel inputs used to power machinery and to manufacture synthetic fertilizers, albeit to a lesser degree. The IPCC estimates that agricultural emissions account for 10–12 per cent of the global total (IPCC, 2007) and that by 2030 agricultural emissions are projected to grow by 36–63 per cent (Smith et al, 2007). For the UK context, emissions associated with agriculture account for nearly half of the food chain's total impacts, or around 8.5 per cent of the UK's total GHG emissions (Garnett, 2008).

Notably, neither the IPCC's estimates, nor those given for the UK, take into account carbon dioxide emissions resulting from agriculturally induced changes in land use (such as deforestation, land degradation and the conversion of pasture to arable) since these are considered in a separate category. However, it is clear that these changes are very significantly driven by agricultural production, particularly of some foods, and these add considerably to farm stage impacts. One global estimate suggests that farming-related land use change emissions add a further 6–17 per cent to the total agricultural burden (Bellarby et al, 2008), putting agriculture's contribution to global emissions at between 16 and 29 per cent of the global total, or as much as 32 per cent once fossil fuel inputs are included. More recently a UK study estimates that if land use change emissions attributable to UK food consumption are included, the contribution of the UK food system to UK emissions rises from around 20 per cent to 30 per cent (Audsley et al, 2010).

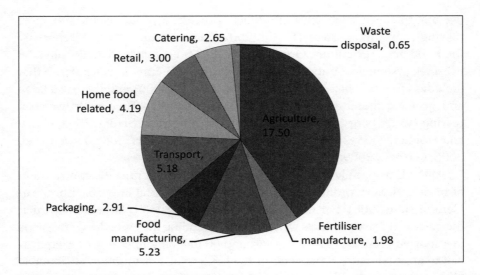

Note: Figure shows UK food-related GHG emissions as a percentage of overall UK consumption-related GHG emissions. A consumption-based food calculation quantifies all emissions produced as a result of a nation's food consumption, including the embedded emissions in imported foods and excluding emissions from those that are exported. These emissions are considered in the context of the UK's total consumption-related GHG emissions, a figure that includes the embedded emissions in all goods imported (from steel to bananas to flip-flops) and that excludes the embedded emissions in products that the country exports. The consumption approach represents a more accurate picture of a nation's contribution to global GHG emissions. See Druckman et al, 2008.

Source: Garnett (2008)

Figure 3.1 Food GHG emissions by life cycle stage

Beyond the farm, the bulk of emissions are attributable to the fossil fuel used to process, transport and retail goods, for refrigeration and for cooking. In the UK, these collectively account for just over half of food's total impacts. Estimates for the developing world have not been undertaken but it is likely that, given the lack of fossil input-using processing and logistics infrastructure, the agricultural stage emissions will, relatively speaking, be more significant. Figure 3.1 illustrates the breakdown of emissions by life cycle stage for the UK.

Greenhouse gas emissions and the contribution of the livestock sector

While the production and consumption of all foods contributes to GHG emissions, it is increasingly recognized that livestock's contribution to the food total is particularly significant. The vast majority of livestock's impacts occur at the farm stage, with subsequent processing, retailing and transport playing more minor roles (Berlin, 2002; Foster et al, 2006). A report by the Food and Agriculture Organization (FAO) estimates that, globally, the livestock system accounts for 18 per cent of GHG emissions – a figure that includes land use change impacts (Steinfeld et al, 2006) and there is a large and growing literature on the GHG emissions associated with livestock rearing (Cederberg and Mattson, 2000; Cederberg and Stadig, 2003; Casey and Holden, 2005, 2006; Basset-Mens and van der Werf, 2005; Lovett et al, 2006; Garnett, 2009; EBLEX, 2009).

At the European level, an EU-commissioned report puts the contribution of meat and dairy products at about half of food's total impacts (European Commission, 2006). For the UK, it has been estimated (Garnett, 2008) that the meat and dairy products we consume (including the embedded emissions in imported products) give rise to around 60 million tonnes of CO_2e, equivalent to approximately 8 per cent[2] of the UK's total consumption attributable GHG emissions or 38 per cent of all food's impacts. Importantly, the figure does not include land use change impacts associated with the production of feedstuffs overseas destined for UK livestock. Livestock account for the bulk of the UK's total agricultural impacts through the methane that ruminants emit, and through the nitrous oxide emitted both directly by livestock and indirectly during the cultivation of feedcrops. According to the FAO, by 2050 the global production and consumption of livestock products is set to double. This clearly has implications for GHG emissions.

Life cycle analysis shows that ruminant animals, such as cattle and sheep, appear to be far more GHG intensive than monogastrics, such as pigs and poultry (see Table 3.1). This is because the former emit methane during the process of digesting their food.

Table 3.1 *GHG intensity of livestock types in the UK*

Livestock type (per tonne of carcass weight, per 20,000 eggs (about 1 tonne) or per 10m³ milk (about 1 tonne dry matter equivalent)	Tonnes CO_2e/tonne
Beef	16.0
Pig (pork & bacon)	6.4
Poultry	4.6
Sheep	17.0
Eggs	5.5
Milk and products	10.6

Source: Williams et al (2006). Overall supply for consumption figures from Defra, 2008a

Methane is not only a potent GHG but its production leads to energy losses, which means that the feed conversion energy of ruminants is lower than that of monogastrics. As a result, the amount of feed energy needed to produce a kilo of beef or of milk solids is greater than that required to produce an equivalent quantity of chicken, pork or eggs. Since the bulk of the anticipated increase in production and consumption will be met by increases in pig and poultry production, with much smaller increases in ruminant production, this is seen to act as a modifying influence in the growth of livestock-related GHG emissions (Steinfeld et al, 2006).

The apparent GHG superiority of white meat to red, and the implications for trends in emissions is, however, complicated by various factors relating to the benefits that livestock bring, as we discuss.

The GHG benefits of livestock production

While livestock production contributes significantly to food-related GHG emissions, to conclude that a vegan agricultural and food system would be preferable is far too simplistic. Livestock farming has been practiced for millennia for good reason – livestock yield multiple benefits, and many of these are environmental. Table 3.2 briefly summarizes the benefits of livestock production and sets them against some of the disbenefits.

For a start it is worth pointing out the obvious: eating will always carry with it an environmental 'cost', although plant based foods, on the whole, generate fewer GHG emissions. Meat and dairy products are an excellent source of nutrition, providing in concentrated form, a range of essential nutrients, including energy, protein, iron, zinc, calcium, vitamin B12 and fat. This said, these nutrients can also be obtained from plant-based foods (or, in the case of vitamin B12, by fortification) (Appleby et al, 1999; Key et al, 1999; Sanders, 1999; Millward, 1999) and the importance of meat and dairy products in supplying these nutrients very much depends on where you are in the world and how rich you are.

On the one hand, in rich societies suffering from the health burdens of over-nutrition, the superabundance particularly of excessive fat and energy in fat- and calorie-rich animal products, can be actively deleterious, while a diverse and nutritionally adequate range of plant based foods are widely available. For these people, plant-based foods can provide adequate nutrition at lower GHG 'cost'.

Among poor societies, however, where meals are overwhelmingly grain or tuber based, where access to a nutritionally varied selection of foods is limited and where there are serious problems of mal- and under-nutrition, keeping a goat, a pig or a few chickens can make a critical difference to the adequacy

Table 3.2 Benefits and disbenefits of livestock production

	Benefits	Disbenefits	Comment
Nutrition	Excellent for protein, calcium, iron, zinc, vitamin B12	Excessive fat; protein can be supplied in excess of requirements	Animal foods not essential, plants can substitute at lower GHG cost
Non-food benefits	Leather, wool, manure, rendered products	Manure can be a pollutant; leather production is a chemical intensive activity	Is supply excess to demand?
Carbon storage	Pasture land stores and (according to some studies) can sequester carbon	Excessive grazing releases carbon; deforestation for pasture land releases carbon	Land use change from pasture to crops releases CO_2
Resource efficiency	Livestock can consume by-products	Grains and cereals feature very heavily in intensive systems	By-products could be used as a feedstock for anaerobic digestion or combusted for heat
Geography	Some land not suitable for cropping	Arable land used to grow cereals and oilseeds for livestock	Uplands and marginal lands could be used for biomass production; agroforestry schemes could incorporate livestock

of the diet (Neumann et al, 2002). Moreover in many communities, livestock play valuable cultural and economic functions; they can be bought or sold as need arises, acting in effect as a form of mobile banking service and so contributing to the food and economic security of the household (Aklilu et al, 2008). Here the relative benefits of animal source foods are considerable.

It is also important to note that animals provide us not only with nutrition but with non-food goods such as leather, manure (which improves soil quality and reduces the need for synthetic fertilizers), traction power and wool. If people did not obtain these goods from livestock, they would need to be produced by some other means and this would almost inevitably incur an environmental cost.

Perhaps most importantly, some livestock systems actively contribute to the avoidance of GHG emissions in two ways. At the right stocking density (and this is critical), grazing animals have an important role to play in maintaining and, in some cases, building carbon stocks in the soil (Allard et al, 2007) – although the evidence on sequestration is mixed. Ploughing this land for arable production would lead to the release of GHG emissions. It is also the case that much of the land used by grazing ruminants is not suited to other forms of food production (for example the Welsh uplands, or the

Mongolian steppes). By eating animals that are reared on land unsuited to other food producing purposes, we avoid the need to plough up alternative land to grow food elsewhere. Note that these potential carbon sequestering, land utilization benefits are obtained from ruminant livestock systems and not from pigs and poultry – a benefit that is not captured by life cycle analysis or in the figures presented in Table 3.1, above. Livestock can also play a vital role in mixed crop rotations; by consuming the clover planted to fertilize the soil they give value (through the milk they provide) to what would otherwise be an economically unproductive part of the rotation. Their manure also helps fertilize and improve the structure of the soil.

Livestock farming has another GHG-avoiding role in that livestock consume by-products that we cannot or will not eat. We, by eating animals who have themselves been fed on food and agricultural by-products, are consuming 'waste' made edible, and in so doing we avoid the need to grow food on alternative land – a process that could give rise to land use change emissions and require fossil fuel based inputs. The feeding of some by-products to livestock is now limited in the EU by legislation; for example, there has been a ban on feeding pigs catering waste, or swill, of mixed plant and animal origin since 2003 (this ban was introduced in 2001 in the UK), following the outbreak of foot and mouth disease. And of course caution is needed: the feeding of certain by-products to animals can (as in the case of BSE) be catastrophic; in many developing world countries pig farming in peri-urban areas, where the pigs are fed on sewage, can give rise to food safety problems.

Moreover, when considering the benefits of livestock for soil carbon sequestration and resource utilization, it should be borne in mind that the gains are highly dependent on the type of system within which the live-stock are reared and the scale of the demand for livestock products. At the extremes both of extensive and intensive livestock systems, the benefits tend to be outweighed by the disbenefits.

In extensive systems (such as those found in many parts of the developing world) that do not use additional feed inputs there can be soil carbon losses that result from overgrazing (Abril and Bucher, 2001) – that is, when the number of cattle being reared is greater than the land's carrying capacity. This is a significant concern in the developing world and it has been esti-mated that 20 per cent of land globally is degraded (Steinfeld et al, 2006). Note that the UK is implicated in overgrazing-related carbon losses over-seas when our demand for major agricultural commodities (often grown to feed intensively reared British livestock), pushes poor livestock farmers on to increasingly marginal and vulnerable pasture lands where soils are quickly degraded. Moreover, extensive cattle farming systems, such as Brazilian ranching, have a highly damaging effect because they trigger a shift from a very high-carbon sequestering form of land use (forest) to one that sequesters less carbon (pasture) – quite aside from the catastrophic impacts on biodiversity. Hence the potential carbon sequestering role of livestock depends on good land management, the maintenance of appropriate stocking densities and – critically – on constraining expansion. In extensive systems where the farm animals are reared in climatically or geologically extreme conditions and receive no supplementation at all, they may also suffer from welfare problems.

Intensive systems, such as those found in the UK, present a different set of problems. For a start, while ruminants do graze on grasslands, these grasslands are not always a 'free' resource. In all, some 66 per cent of the grassland area in the UK receives nitrogen fertilizer applications (Defra, 2007), and these give rise to N_2O emissions. While sheep and some cattle are indeed left for much of their lives to graze on the uplands, they are usually finished on fertilized lowland grass, perhaps supplemented with concentrates – without this extra input, the meat yield would be too low to be economically viable. Moreover, livestock of all types in intensive systems not only consume by-products but also large quantities of cereals that, arguably, could have been eaten directly and more efficiently by humans. Measured in terms of land area or GHG emissions per unit of protein or calories, it is less efficient to feed grain to animals that we then eat, than it is for us to eat the grain directly. Globally, livestock have been estimated to consume over a third of world cereal output (FAO, 2009b) and the proportion is higher still in the UK (Defra, 2008a). The use of land to produce cereals for livestock ultimately leads to land use change either directly, by colonizing new land, or indirectly, by pushing existing activities (such as pastoral grazing) into new land.

The negative impacts of soy, a major input to intensive systems, are particularly striking. While soy-oil has its uses in industrial food manufacture and increasingly as a biofuel, the cake accounts for around two thirds of the crop's economic value (since it is produced in larger quantities than the oil) (FAO, 2009b) and as such in some years can actually drive soy production. Soy cannot, then, accurately be called a 'by-product'.

Soy farming has major implications for GHG emissions because of its role in land use change and particularly in deforestation.(WWF, 2004; Nepstad et al, 2006; McAlpine et al, 2009). While cattle ranching is the major driver of deforestation in Brazilian Amazonia, accounting for the bulk of direct deforestation, the relationship between cattle ranching and other drivers, particularly soy, is both close and complex. Cattle ranches are often set up to secure land tenure and to maintain cleared land, allowing other more profitable enterprises such as soy to move in (McAlpine et al, 2009). In the decade up to 2004, industrial soybean farming doubled its area to $22,000km^2$ and is now the largest arable land user in Brazil (Elferink et al, 2007). Moreover, soybean cultivation not only makes use of land in its own right, but is also an important 'push' factor for deforestation by other industries; it takes land away from other uses, such as smallholder cultivation and cattle rearing, and pushes these enterprises into the rainforest (Nepstad et al, 2006; Fearnside and Hall-Beyer, 2007). Additionally, it provides income to purchase land for other purposes, including logging. Hence while beef cattle are a major direct cause of Brazilian deforestation, the role of soy as a second stage colonizer is significant.

As highlighted earlier, the bulk of the projected doubling in the production and consumption of livestock is due to increases in intensive pig and poultry production. Intensively reared pigs and poultry are major consumers of soy. Thirty-two per cent of Brazil's soy feed was exported to the European Union in 2006/7 and 90 per cent of that soy went to feed Europe's pigs and poultry (Friends of the Earth Netherlands, 2008); this 'hidden' land use cost should properly be added to the emissions shown in Table 3.1.

Moreover, while pigs and poultry in intensive systems may be efficient converters of feed into meat, they rely on cereals and soy – in fact actively compete, through the demand for grains, with land needed to grow crops for humans. As such, farming them in this way is implicated in land use change. Unlike grazing animals, they do not provide the benefit of carbon storage services.

This said, if the projected increase in animal source foods were met by either by increases in extensive or intensive ruminant farming the problem could arguably be worse. The direct land needs of ruminant cattle are large and expansion into forest areas would lead to major soil carbon losses. As for intensive beef and dairy systems – again, set to grow – these require not only grazing land but also significant cereal and oilseed inputs, which, as noted, they consume less efficiently than do pigs and poultry. As such, high-volume, intensive ruminant production would represent the worst of both worlds. We would have a system that emits methane (as well as the nitrous oxide that all animals emit); and that is also dependent on cereals and soy – a double whammy. Expansion of extensive grass fed systems does not, therefore present a solution.

To conclude, ruminant livestock *can* yield carbon storage and resource efficiency benefits but this only takes place in certain extensive systems at certain stocking densities. Once the numbers of livestock are greater than the ability of the land to support them, then the disbenefits, in the form of land degradation (and lower animal welfare) rise to the fore. Similarly, while livestock farming of all types can represent a form of resource efficiency, the volume of by-products available cannot alone feed the sheer numbers of animals that we appear to want to eat. Growth in demand is what turns a sustainable system into an unsustainable system.

Hence what emerges when discussing the benefits and disbenefits of livestock systems is that two factors are critical: the *type* of livestock system, and the *scale* of livestock production and associated consumption. The right type of system at the right scale is needed to ensure that the benefits outweigh the disbenefits.

Reducing livestock-related GHG emissions: What can be done?

Policy makers in the developed world are increasingly aware of the need to tackle livestock-related emissions. There is now a very considerable and growing body of research examining how GHG emissions from agriculture in general, and from livestock more specifically, might be reduced (Clark et al, 2001; Committee on Climate Change, 2008; Defra, 2008b; Garnett, 2008; Smith et al, 2008).

Broadly speaking, the focus is on efficiency: an approach that takes as its starting point the need to meet growing demand at lowest GHG cost. As highlighted, the key word here is demand. It is projected – and accepted – that demand for meat and dairy products will double by 2050, and while the consumption issue is starting to receive some attention in the developed world, growth in the developing world is taken to be inevitable.

Efficiency is taken to mean the production of as much meat, milk and associated products for as little environmental and land use 'cost' as possible. We explore the extent to which intensive production achieves its goals and highlight the implications for animal welfare, biodiversity and human health, before setting out an alternative approach.

The efficiency approach

The research literature on livestock GHG mitigation tends to focus on four main categories of action: improving the efficiency of breeding and feeding strategies; managing soils to sequester carbon; managing manure to reduce methane, nitrous oxide and ammonia emissions and (through anaerobic digestion) to produce energy; and decarbonizing energy inputs. Table 3.3 sets these out in more detail.

Approaches b and c relate to agriculture in general. Approach a is specifically livestock related and encompasses a range of strategies. A key element is to manage the feeding regime to achieve an 'optimal' balance between carbohydrate and protein – this is to ensure that the food is used by the animal to produce milk or meat rather than being 'wasted' in the form of methane or nitrogen losses. Note that the emphasis is very much on the use of cereal and protein inputs; ruminants fed a mixed diet of cereals and proteins emit fewer methane emissions per unit of milk or meat than

Table 3.3 *GHG mitigation – an efficiency approach*

Approach	Goals	Example measures
a **Efficient husbandry**	More milk or meat per kg of GHG emissions	Optimizing feed conversion ratio through use of cereals and oilseeds in the right proportions to maximize yields at minimum GHG cost.
		Breeding for better partitioning of feed into milk (dairy); for better muscle tone (beef); for rapid growth (pigs and poultry)
		Methane inhibitors, feed supplements (e.g. oils), bovine somatotropin, genetic modification (all still at experimental stage) Breeding for fertility and longevity (dairy cows)
b **Soil carbon management**	Retain or increase carbon in the soil	Improving pasture quality; restoring degraded lands; minimum and no till arable farming (feed and food crops)
c **Manure management**	Reduce methane, nitrous oxide and ammonia emissions; create biogas as a substitute for fossil fuels	Anaerobic digestion to produce biogas to substitute for fossil energy and an anaerobic digestate as a substitute for fossil fuel based fertilizer inputs; composting manure to stabilize and improve its quality
d **Energy decarbonization**	Reduce CO_2 from fossil energy inputs	Anaerobic digestion as on farm energy source; more efficient energy management; renewables; reduced use of synthetic fertilizers

grass-fed cattle (since there is less roughage) and breeding strategies are designed to maximize the capacity of the animal to be productive on a diet of concentrates. Similarly poultry and pig systems based on the rearing of fast-growing animals fed on energy and protein intensive concentrates reach slaughter weight rapidly. Emissions are therefore lower, since fewer days need to be spent heating and lighting their housing, and the feed conversion efficiency is higher.

The feed-optimization approach can be combined, in the case of dairy cows, with breeding strategies to increase productivity (by increasing that portion of feed intakes that is partitioned, through metabolic processes into milk) and fertility. To date there has been a strong emphasis on breeding for increased productivity, at the cost of reduced fertility, but there are signs now of greater focus on extending fertility. The fertility issue is relevant from a GHG, as well as from an animal welfare perspective, because a rapid turnover of milkers due to early death or infertility means that energy inputs and greenhouse gas outputs are 'wasted' in the process of rearing heifers before they reach their first pregnancy and lactation. Once she has reached maturity, it important to keep the cow milking for as long as possible after this period so that the investment in growth and development pays off and to keep the replacement rate as low as possible.

Parallel strategies for feed production include increasing the productivity of feed crops so that more crops can be grown on a given area of land, again through breeding and fertilizer strategies. Biomass production could form part of the efficiency picture. As rearing livestock on uplands becomes increasingly unprofitable, livestock farmers are leaving the hills – a trend we are witnessing now as a result of the Common Agricultural Policy. An alternative use for the uplands might be biomass production, an approach that contributes to carbon sequestration, and also generates a fuel source. Under an efficiency scenario, most of the growth in livestock products will come from more profitable pigs and poultry, with their higher feed conversion efficiencies.

Note that these feeding and breeding approaches, while 'efficient', are inherently dependent on cereals and oilseeds. More grains to feed more animals will mean changes in land use and the CO_2 impacts of this are not, as yet, taken into account in the life cycle analyses of the sort that produces the figures in Table 3.1.

Some attempts have been made to quantify the potential GHG savings achievable through a combination of approaches a–d. Most estimates consider the agricultural sector as a whole, as it is difficult to separate out livestock farming given its interconnectedness with the arable sector. However, since livestock contribute to the bulk of agricultural emissions (either through their direct emissions or through emissions generated by the production of feed crops for their consumption), these estimates are indicative of what might be possible for livestock systems too.

It is generally agreed that, for the UK, reductions of up to 30 per cent for agriculture may be technically possible, although what is economically and politically feasible will be significantly lower (Committee on Climate Change, 2008; Defra, 2008b; Garnett, 2008). There do not appear to be any estimates of the potential specifically in developing world countries but some attempts at a global estimate have been made. The IPCC suggests that by 2030 mitigation measures – largely soil carbon management (b, above) – could offset by 70–80 per cent of *today's* direct emissions. However, it points out that under a business-as-usual scenario, agricultural emissions as a whole are set to grow by 36–63 per cent (although there are huge uncertainties), due to increases in demand for food in general and for animal source foods in particular. As such, while there may be reductions on a per kg of food basis, there may be no absolute reduction in emissions (Schneider et al, 2007). It also points out

Table 3.4 *Meat and dairy demand in 2000 and predicted demand in 2050*

	2000 *(population 6 billion)*	2050 *(population 9 billion)*
Average per capita annual global demand – meat (tonnes)	0.0374	0.052
Average per capita annual global demand – milk (tonnes)	0.0783	0.115
Total annual demand – meat (million tonnes)	228	459
Total annual demand – milk (million tonnes)	475	883

Source: FAO (2006)

that 90 per cent of the emission reductions modelled come from soil carbon management and that these practices are time limited – once the soil has reached its maximum capacity to accumulate carbon, no further sequestration arises. Hence, by 2050, there will be far less to be gained from the soil carbon management approach. As such, it becomes increasingly important to tackle nitrous oxide and methane emissions, and their relative importance is set to grow. Note that the IPCC estimates of the mitigation potential do not include those arising from land use change, which, as noted, above, contribute to an estimated 6–17 per cent of global emissions.

The general picture that emerges, amidst enormous uncertainty, is that technical and managerial approaches are not enough. The 50 per cent reduction in emissions per unit of meat or milk will be offset by the doubling in demand. We need, however, to reduce emissions absolutely by 50 per cent by 2050. If the food chain does not play its part in contributing to the emissions reduction, then other sectors of society (housing, transport, energy supply and so forth) will have to reduce their emissions even more to compensate; and these sectors, given a growing global population, face exactly the same sort of challenges as does the food chain.

Since technology cannot get us where we need to be, we need to look at changing the balance of foods we consume – and this will include reducing our consumption of meat and dairy products.

Meat and dairy intakes are particularly high in the developed world and historically the developed world is responsible for the bulk of GHG emissions. One equitable approach to consider, then, would be to examine what might happen if people in rich societies were to reduce their consumption of animal products. One possible option would be for the world's population to converge on consuming what in 2050 people in the developing world are anticipated to consume: about 44kg of meat and 78kg of milk annually.

This represents a 62 per cent and 73 per cent increase on average meat and milk consumption in developing countries today, although this average masks very wide inequities in distribution: average per capita annual meat and milk consumption in Ethiopa, for example, is 8kg and 21kg respectively. For people in the developed world, however, consuming at this level would entail a very substantial change in habits. It would mean that we in the UK would halve the amount of meat we typically eat today, and reduce our milk consumption by an even more drastic two thirds.

Unfortunately, however, action by the developed world alone will not be sufficient simply because the bulk of the projected increase in demand is set to come from the developing world. If we multiply reductions in per capita consumption by the number of people who are projected to be living in the developed and transition countries, and subtract this figure from the overall *anticipated* demand for meat and dairy products, we obtain a mere 15 per cent overall reduction in *projected* world meat consumption, and 22 per cent for milk, as Table 3.5 shows. Overall global volumes will still be higher than

Table 3.5 *Reduction achieved by developed world only reduction in consumption*

Meat	Population 2050, bn	Projected T/person/ yr 2050	Total anticipated consumption mill T 2050	Total consumption all at 2050 developing world levels mill T	% reduction in consumption compared with BAU projections
Developed countries	1.019	0.103	105	44. 8	
Developing countries	7.51	0.044	330.4	330.4	
Transition countries	0.343	0.068	23.3	151	
World	8.92		458.7		15
Milk					
Developed countries	1.019	0.227	231.3	794.8	
Developing countries	7.51	0.078	585.7	585.7	
Transition countries	0.343	0.193	66.2	267.5	
World	8.92		883.2		22

Note: Transition economies are those former socialist countries of the USSR and Eastern Europe.

Source: Based on data presented in 'Prospects for Food, Nutrition, Agriculture and Major Commodity Groups. World Agriculture: Towards 2030/2050', Interim report, Global Perspective Studies Unit, Food and Agriculture Organization of the United Nations, Rome, June 2006

they are today; 70 per cent higher for meat, and 45 per cent higher for milk. Clearly, reductions at this level are not sufficient. Another approach is to ask how much would be available to each individual in 2050 if we keep meat and dairy production at 2000 levels, so as to avoid a rise in livestock-related GHG emissions. A very simple calculation finds that in the context of 9 billion people in 2050, per capita consumption of meat and milk would need to be as low as 25kg and 53kg a year respectively. This is approximately the average level of consumption of people in the developing world *today* and equates to half a kilo of meat and a litre of milk per person per week.

These figures are strikingly low – they imply drastic declines for the rich and allow for no increase by the poor. Diets low in animal products can be nutritionally adequate but – as highlighted – much depends on what else there is to eat and how equitably it is distributed. Policy makers need to think about developing food and agricultural strategies based on combining food and climate change policies – of prioritizing food security at minimum greenhouse gas cost.

The efficiency mindset as characterized by approach a, above, is not only unable to meet our emission goals but it also considers animals and their emissions simplistically as a problem to be managed in order to meet demand. Issues such as animal welfare, the health implications of increased meat and dairy consumption for the 1 billion of the world's population who are overweight or obese (WHO, n.d.) and biodiversity losses are to be addressed by other means. Examples include marketing niche higher welfare systems for 'premium' customers who demand it, developing (for obesity) low fat food formulations, functional foods, nutraceuticals and exercise programmes and intensifying production on agricultural land so as to create biodiversity havens, or ghettos on the land that remains.

There is an alternative way of thinking, however, which could be characterized by the phrase 'livestock on leftovers'. Such an approach seeks to work with what animals are good at – making use of marginal land, and feeding on by-products that we cannot eat. It considers what the land and available by-products can sustainably support – and then assesses, on that basis, how much is available for us to eat. It takes land and the biodiversity that it supports as its starting point, and as its ultimate constraint. This mindset considers demand to be negotiable and challengeable.

With this approach livestock can be integrated into a landscape so that they help store carbon in the soil, enhance the biodiversity of local ecosystems and make use of the leftovers from other food and agricultural processes. There will be a need to focus research on breeding programmes that emphasize robustness and flexibility; that improve the ability of ruminants to survive on marginal lands and of all livestock types to respond well to a variable supply of different foods – to cope well with a less nutritionally precise environment, as it were. We need also to consider how livestock can be reintegrated into arable farming systems in the developed world, as part of a mixed livestock–crop rotations and livestock–agroforestry systems.

A 'livestock on leftovers' approach need not, and should not, be purist. For example, in many developing world countries, 'intensification' may include actions to ensure that cows have something better to eat than plastic bags – and there will be welfare as well as productivity gains from so doing. Options b, c and d outlined in Table 3.3 above will still be key to overall agricultural mitigation.

This 'livestock on leftovers' approach is likely to provide us with far lower quantities of meat and dairy than that afforded by intensification – but the system actively helps deal with the problem of climate change, while the efficiency approach is simply geared towards minimizing the damage that livestock cause. A livestock on leftovers approach would form part of an agricultural strategy that takes as its starting point the need to meet nutritional needs, while mitigating agricultural emissions.

A short note on fish

This chapter has focused on only on terrestrial animal source foods and as such has omitted a very significant contributor to human protein intakes: fish. While the volume of fish from capture fisheries has remained fairly stable, aquaculture has been growing at about 8 per cent a year and now accounts for about a third of all fish harvested (and around half of fish destined directly for human consumption) (FAO, 2008). There is a risk that if people reduce their consumption of meat and dairy products, they will increase the amount of fish they eat.

While from a GHG perspective, fish on the whole have a lower GHG footprint than terrestrial livestock, this generalization masks wide variation in the relative intensities of different fish species. For example a study by Tyedmers into fuel use in fishing reported a range of fuel inputs to fisheries of 20–2000 litres of fuel per tonne of fish landed, largely becuase of differences in the intensity of fuel use by fishing vessels (Tyedmers, 2004). There are similarly wide differences in aquaculture; the variation in intensity here depends in part on the feed source (particularly the amount of fuel used in the fishing of the wild fish that go to feed the farmed fish) and partly on the inherent feed conversion efficiency of the farmed fish themselves (Pelletier and Tyedmers, 2010).

It is also very important to emphasize that a simplistic carbon accounting assessment risks ignoring the very serious broader environmental and ecosystemic issues associated with the fish and aquaculture sectors. In the case of wild fish, just of half of all stocks that have been monitored are now fully exploited; over a quarter are over-exploited, depleted or slowly recovering, with the remainder either under-exploited or moderately exploited (FAO, 2008).

As regards aquaculture, this is associated in some parts of the world with significant environmental problems, including coastal pollution, the destruction of mangrove swamps and the escape of farmed fish into the seas, spreading disease. Carnivorous fish, such as salmon, trout and tuna and omnivores like shrimps that are often reared on a carnivorous diet, are fed wild fish from stocks that may themselves be depleted – and these kinds of fish are the species that have grown most rapidly in populariy.

On the other hand, many forms of aquaculture – including all plant (such as seaweed) and mollusc cultivation, polyculture systems that incorporate these species and many forms of extensive to semi-intensive freshwater finfish production – rely little on marine inputs such as fish meal and could potentially make a significant and sustainable contribution to global food protein supplies (Tyedmers et al, 2007). There is in particular, as yet, unexplored scope for increasing the use of algae and seaweeds, both as feedstuffs and for consumption in their own right.

In short, the picture currently presented by aquaculture is mixed but there is great potential for improving the sustainability of the aquaculture sector (in particular in reducing its dependence on wild fish as a feed-source) and, in so doing, providing an alternative to livestock consumption and production.

Conclusions

Our food system faces enormous and difficult challenges. We need to halve food and agricultural emissions by 2050 while feeding a global population that will be a third higher than it is today. We also need to meet these goals within the constraints of what are essentially ethical 'non- negotiables': the safeguarding of biodiversity and a decent quality of life for the animals that we rear and eat. We are bound, of course, by the ultimate constraint – land.

The projected doubling in demand for meat and dairy products presents a possibly insuperable obstacle. At present technological improvements will not allow us to meet this demand and at the same time reduce emissions to the degree needed. It may conceivably be that we achieve a technological breakthrough, enabling us to meet demand while also reducing emissions – but it is likely that this will come at the expense of animal welfare and of biodiversity. There is, moreover, no guarantee that by producing enough food we achieve food security. Distribution and access are socio-economic, not just biological, challenges. One might argue that a more redistributive approach to meeting the food needs of the most vulnerable will be mindful of the environmental impacts – since it is the poorest who have to live most directly with the consequences of climate change.

By contrast, a business-as-usual approach continues the global trend towards further dependence on energy- and GHG-intensive lifestyles, and the challenge of trying to meet these demands will continue. By 2050, on current projections, the developing world will still, on average, be eating less than half as much meat as people do in the rich world, and only a third of the milk. There is a long way to go before they catch up with developed world levels. Do we assume that ultimately they will want to eat as much meat and milk as we do, and do markets therefore seek to supply these volumes? When is enough enough? Who decides at what level justifiable wants turn into unsustainable greed? We need to start questioning the unquestionable – demand.

Time is running out. We have little time left to avert the worst impacts of climate change. We need to start tackling the problems we face – food security, climate change, animal welfare, biodiversity – in an integrated way rather than through the separate, sometimes conflicting strategies that we have today. The vision should be to achieve good nutrition for all at minimum environmental cost, and global policy makers will need to develop fiscal, regulatory and other measures to make this happen. Farm animals can play an important part in achieving this vision, but to do so, the livestock sector needs to understand and work with the strengths and limits of the land and its resources.

Notes

1   The IPCC defines this as 'over 90% certainty'.
2   Note that the figure is almost the same as the total attributable to agriculture. This results from different methods of quantifying agricultural and livestock emissions and different data sources and different boundaries. However, what is clear (and this is also evident from the UK GHG inventory) is that livestock account for the bulk of agricultural GHG emissions.

References

Abril, A. and Bucher, E. H. (2001) 'Overgrazing and soil carbon dynamics in the western Chaco of Argentina', *Applied Soil Ecology*, vol 16, no 3, pp243–249

Aklilu, H. A., Udo, H. M. J., Almekinders, C. J. M. and Van der Zijpp, A. J. (2008) 'How resource poor households value and access poultry: Village poultry keeping in Tigray, Ethiopia', *Agricultural Systems*, vol 96, pp175–183

Allard, V., Soussana, J.-F., Falcimagne, R., Berbigier, P., Bonnefond, J. M., Ceschia, E., D'hour, P., Hénault, C., Laville, P., Martin, C. and Pinarès-Patino, C. (2007) 'The role of grazing management for the net biome productivity and greenhouse gas budget (CO_2, N_2O and CH_4) of semi-natural grassland', *Agriculture, Ecosystems and Environment*, vol 121, pp47–58

Anderson, K. and Bows, A. (2008) 'Reframing the climate change challenge in light of post-2000 emission trends', *Philosophical Transactions of the Royal Society A*, vol 366, no 1882, pp3863–3682

Appleby, P. N., Thorogood, M., Mann, J. I. and Key, T. J. (1999) 'The Oxford Vegetarian Study: An overview', *American Journal of Clinical Nutrition*, vol. 70 (Supplement 3), pp525S–531S

Audsley, E., Brander, M., Chatterton, J., Murphy-Bokern, D., Webster, C. and Williams, A. (2010) *How Low Can We Go? An Assessment of Greenhouse Gas Emissions from the UK food System and the Scope to Reduce Them by 2050*, WWF-UK, Godalming

Basset-Mens, C. and van der Werf, H. M. G. (2005) 'Scenario-based environmental assessment of farming systems: The case of pig production in France', *Agriculture, Ecosystems and Environment*, vol 105, pp127–144

Bellarby, J., Foereid, B., Hastings, A. and Smith, P. (2008) 'Cool farming: Climate impacts of agriculture and mitigation potential', report produced by the University of Aberdeen for Greenpeace

Berlin, J. (2002) 'Environmental life cycle assessment (LCA) of Swedish semi-hard cheese', *International Dairy Journal*, vol 12, pp939–953

Casey, J. W. and Holden, N. M. (2005) 'The relationship between greenhouse gas emissions and the intensity of milk production in Ireland', *Journal of Environmental Quality*, vol 34, pp429–436

Casey, J. W. and Holden, N. M. (2006) 'Quantification of greenhouse gas emissions from suckler-beef production in Ireland', *Agricultural Systems*, vol 90, pp79–98

Cederberg, C. and Mattson, B. (2000) 'Life cycle assessment of milk production – a comparison of conventional and organic farming', *Journal of Cleaner Production*, vol 8, pp49–60

Cederberg, C. and Stadig, M. (2003) 'System expansion and allocation in life cycle assessment of milk and beef production', *International Journal of Lifecycle Assessment*, vol 8, no 6, pp350–356

Clark, H., de Klein, C. and Newton, P. (2001) 'Potential management practices and technologies to reduce nitrous oxide, methane and carbon dioxide emissions from New Zealand agriculture', prepared for Ministry of Agriculture and Forestry, New Zealand, September 2001

Committee on Climate Change (2008) 'Building a low-carbon economy – the UK's contribution to tackling climate change', The First Report of the Committee on Climate Change, The Stationery Office, London, UK

Defra (2007) Table B.10, 'The British survey of fertiliser practice: Fertiliser use on farm crops for crop year 2007', Defra, UK

Defra (2008a) 'Agriculture in the United Kingdom', www.defra.gov.uk/evidence/statistics/foodfarm/general/auk/documents/AUK-2008.pdf, last accessed 16 July 2010

Defra (2008b) 'The milk road map', produced by the Dairy Supply Chain Forum's Sustainable Consumption & Production Taskforce, Defra, UK, May 2008

Diamond, J. (2005) *Collapse*, Penguin, London

Druckman, A., Bradley, P. and Papathanasopoulou, E. (2008) 'Measuring progress towards carbon reduction in the UK', *Ecological Economics*, vol 66, pp594–604

English Beef and Lamb Executive (EBLEX) (2009) *Change in the Air: The English Sheep and Beef Production Roadmap – Phase One*, EBLEX, Warwickshire, UK

Elferink, E. V., Nonhebel, S. and Schoot Uiterkamp, A. J. M. (2007) 'Does the Amazon suffer from BSE prevention?' *Agriculture, Ecosystems and Environment*, vol 120, pp467–469

European Commission (2006) 'Environmental impact of products (EIPRO): Analysis of the life cycle environmental impacts related to the total final consumption of the EU25', European Commission Technical Report EUR 22284 EN, May 2006

FAO (2008) 'The state of world fisheries and aquaculture 2008', Food and Agriculture Organization, Rome

FAO (2009a) 'News release: 1.02 billion people hungry', 19 June 2009, available at www.fao.org/news/story/en/item/20568/icode/, last accessed 24 May 2010

FAO (2009b) 'Food outlook 2009', Food and Agriculture Organization, Rome

FAO (2009c) 'Appendix Table A24 – Selected international prices for oilcrop products and price indices', in 'Food Outlook 2009', Food and Agriculture Organization, Rome

Fearnside, P. and Hall-Beyer, M. (2007) 'Deforestation in Amazonia', in Cleveland, C. (ed) *Encyclopedia of Earth*, Environmental Information Coalition, National Council for Science and the Environment, Washington, DC (first published in the *Encyclopedia of Earth* 15 March 2007; last revised 30 March 2007), available at www.eoearth.org/article/Deforestation_in_Amazonia

Foster, C., Green, K., Bleda, M., Dewick, P., Evans, B., Flynn, A. and Mylan, J. (2006) 'Environmental impacts of food production and consumption', report produced for the Department for Environment, Food and Rural Affairs

Friends of the Earth Netherlands (2008) 'Soy consumption for feed and fuel in the European Union', research paper prepared for Milieudefensie (Friends of the Earth Netherlands) by Profundo Economic Research, The Netherlands

Garnett, T. (2008) *Cooking up a Storm: Food, Greenhouse Gas Emissions and Our Changing Climate*, Food Climate Research Network, Centre for Environmental Strategy, University of Surrey, UK

Garnett, T. (2009) 'Livestock-related greenhouse gas emissions: Impacts and options for policy makers', *Environmental Science and Policy*, vol 12, pp491–503

Gupta, S., Tirpak, D. A., Burger, N., Gupta, J., Höhne, N., Boncheva, A. I., Kanoan, G. M., Kolstad, C., Kruger, J. A., Michaelowa, A., Murase, S., Pershing, J., Saijo, T. and Sari, A. (2007) 'Policies, instruments and co-operative arrangements', in Metz, B., Davidson, O. R., Bosch, P. R., Dave, R. and Meyer, L. A. (eds) *Climate Change 2007: Mitigation. Contribution of Working Group III to the Fourth Assessment Report of the Intergovernmental Panel on Climate Change*, Cambridge University Press, Cambridge and New York

Hadley Centre (2005) 'Avoiding dangerous climate change', International Symposium on the Stabilisation of Greenhouse Gas Concentrations, Met Office, Exeter, UK, 1–3 February 2005

IPCC (2007) *Climate Change 2007: Mitigation. Contribution of Working Group III to the Fourth Assessment Report of the Intergovernmental Panel on Climate Change* (Metz, B., Davidson, O. R., Bosch, P. R., Dave, R., Meyer, L. A. (eds)), Cambridge University Press, Cambridge and New York, Chapter 8

Key, T. J. et al (1999) 'Health benefits of a vegetarian diet', *Proceedings of the Nutrition Society*, vol 58, pp271–275

Lovett, D. K., Shalloo, L., Dillon, P. and O'Mara, F. P. (2006) 'A systems approach to quantify greenhouse gas fluxes from pastoral dairy production as affected by management regime', *Agricultural Systems*, vol 88

McAlpine, C. A., Etter, A., Fearnside, P. M., Seabrook, L. and Laurance, W. F. (2009) 'Increasing world consumption of beef as a driver of regional and global change: A call for policy action based on evidence from Queensland (Australia), Colombia and Brazil', *Global Environmental Change*, vol 19, pp21–33

Millward, D. J. (1999) 'The nutritional value of plant based diets in relation to human amino acid and protein requirements', *Proceedings of the Nutrition Society*, vol 58, pp249–260

Nepstad, D. C., Stickler, C. M. and Almeida, O. T. (2006) 'Globalization of the Amazon soy and beef industries: Opportunities for conservation', *Conservation Biology*, vol 20, no 6, pp1595–1603

Neumann, C., Harris, D. M. and Rogers, L. M. (2002) 'Contribution of animal source foods in improving diet quality and function in children in the developing world', *Nutrition Research*, vol 22, nos 1–2, pp193–220

Pelletier, N. and Tyedmers, P. (2010) 'A life cycle assessment of frozen Indonesian tilapia fillets from lake and pond-based production systems', *Journal of Industrial Ecology*, vol 14, no 3, pp467–481

Sanders, T. A. (1999) 'The nutritional adequacy of plant-based diets', *Proceedings of the Nutrition Society*, vol 58, pp265–269

Schellnhuber, H. J., Cramer, W., Nakicenovic, N., Wigley, T. and Yohe, G. (2006) *Avoiding Dangerous Climate Change*, Cambridge University Press, Cambridge

Schneider, S. H., Semenov, S., Patwardhan, A., Burton, I., Magadza, C. H. D., Oppenheimer, M., Pittock, A. B., Rahman, A., Smith, J. B., Suarez, A. and Yamin, F. (2007) 'Assessing key vulnerabilities and the risk from climate change', in Parry, M. L., Canziani, O. F., Palutikof, J. P., van der Linden, P. J. and Hanson, C. E. (eds) *Climate Change 2007: Impacts, Adaptation and Vulnerability. Contribution of Working Group II to the Fourth Assessment Report of the Intergovernmental Panel on Climate Change*, Cambridge University Press, Cambridge, pp779–810

Smith, P., Martino, D., Cai, Z., Gwary, D., Janzen, H. and Kumar P et al (2007) 'Agriculture', in Metz, B., Davidson, O. R., Bosch, P. R., Dave, R., Meyer, L. A. (eds) *Climate Change 2007: Mitigation. Contribution of Working Group III to the Fourth Assessment Report of the Intergovernmental Panel on Climate Change*, Cambridge University Press, New York

Smith, P., Martino, D., Cai, Z., Gwary, D., Janzen, H. H., Kumar, P., McCarl, B., Ogle, S., O'Mara, F., Rice, C., Scholes, R. J., Sirotenko, O., Howden, M., McAllister, T., Pan, G., Romanenkov, V., Schneider, U., Towprayoon, S., Wattenbach, M. and Smith, J. U. (2008) 'Greenhouse gas mitigation in agriculture', *Philosophical Transactions of the Royal Society*, vol 262, pp780–813

Steinfeld, H., Gerber, P., Wassenaar, T., Castel, V., Rosales, M. and de Haan, C. (2006) 'Livestock's long shadow: Environmental issues and options', Food and Agriculture Organization, Rome

Stern, N. (2006) *The Economics of Climate Change: The Stern Review*, Cambridge University Press, Cambridge (Although the Stern review takes as its threshold the higher CO_2e level of 550ppm, the adequacy of this figure has been increasingly called into question and is currently the subject of UK Government scrutiny)

Thomas, K. (1991) *Man and the Natural World: Changing Attitudes in England 1500–1800*, Penguin, London

Tyedmers, P. (2004) 'Fisheries and energy use', in Cleveland, C. (ed) *Encyclopedia of Energy*, vol 2, Elsevier, Amsterdam, pp683–693

Tyedmers, P., Pelletier, N. and Ayer, N. (2007) 'Biophysical sustainability and approaches to marine aquaculture: Development policy in the United States', report to the Marine Aquaculture Task Force, available at www.whoi.edu/sites/marineaquataskforce, last accessed 24 May 2010

United Nations (2008) 'The Millennium Development Goals Report 2008', United Nations, New York

WHO (n.d.) *Obesity and Overweight Factsheet*, World Health Organisation, available at www.who.int/dietphysicalactivity/publications/facts/obesity/en/, last accessed 24 May 2010

Williams, A. G., Audsley, E. and Sandars, D. L. (2006) 'Determining the environmental burdens and resource use in the production of agricultural and horticultural commodities', Main Report, Defra Research Project ISO205, Bedford, Cranfield University and Defra

WWF (2004) 'ISTA Mielke', Oil World Annual 2004, Hamburg, May 2004 cited in Jan Maarten Dros, *Managing the Soy Boom: Two Scenarios of Soy Production Expansion in South America*, available at www.panda.org/downloads/forests/managingthesoyboomenglish_nbvt.pdf, last accessed 24 May 2010

4

Industrial Livestock Production and Biodiversity

Susanne Gura

While in developing countries local breeds are part of everyday life, only a few dedicated people in the industrialized world keep rare breeds. Others fight against plans to build factory farms in their region because of the expected pollution of air, water and soil, and to prevent the destruction of ecosystems. And many more will buy milk, eggs, fish and meat only from organic systems, if at all. They worry about uniformly bred animals kept in unfair conditions on factory farms. All of them want, among other things, to preserve and enjoy biodiversity.

Biological diversity – or biodiversity – has three components:

1   The diversity of plant and animal species, as well as micro-organisms – this is what makes a salmon, a moth, an *E. coli* bacterium, a daffodil or a gorilla different from a human being and from each other.
2   Diversity within a species – this is what makes a person look and be so different from another person. All people belong to the species *Homo sapiens sapientissimus*; similar variety exists within each species.
3   The diversity of ecosystems – without their ecosystems, the species can't live and develop.

This chapter will show how industrial livestock keeping is destroying ecosystems, how it has caused the loss of breeds that is so dangerous in our fragile industrialized food economy and how far genetic uniformity in industrial livestock keeping has come, and is further bound to proceed if plans for cloning and genetic modification succeed.

Slaughtering the Amazon for cattle and livestock feed

According to the Intergovernmental Panel on Climate Change, about a fifth of the Amazon rainforest has already been destroyed, and 80 per cent of this destruction was to make place for cattle pastures. The meat is predominantly consumed in South America. The leather, however, is exported to China, Vietnam and Italy, where leading shoe manufacturers Adidas, Reebok, Nike, Clarks, Geox and Timberland are getting shoes and other leather items produced for the world markets, according to a Greenpeace report presented during the June 2009 climate negotiations in Bonn. Greenpeace asked the shoe producers not to buy leather produced from cattle on farms that are involved in rainforest destruction. Brazil's largest leather producer and second largest meat producer, Bertin, agreed. Its suppliers will provide global positioning system (GPS) data of their farms and a traceability system will prevent the hiding of suppliers (Greenpeace International, 2009).

So far, little has slowed the destruction of the Amazon rainforest. After pastures, the next crop succession step is usually soya beans meant for feeding the world's industrial cattle, pigs, poultry and fish. Indeed, feed crops make up one fifth of former rainforest areas. Europe imports around three quarters of its feed needs, mostly from South America.

While for a long time the general public was mostly worried about the huge biodiversity losses associated with rainforest destruction, the Amazon being home to one tenth of known (and innumerable unknown) species, climate change is currently the overriding issue. Soon, public discussion may have to turn to world hunger. The number of undernourished people has grown from 850 million to more than 1 billion, and is expected to increase further with the pressures from growing bio-energy production on land and water. Calls for lower meat consumption to make it possible to feed the world's population were made, but not acted upon, back in the 1970s. Ecosystem destruction is at a point where many of our youth have lost hope in their future. Are we now ready to reduce meat consumption?

The Amazon rainforest is not the only ecosystem being sacrificed to the livestock industry. The Northern Gulf of Mexico and the Mississippi River basin are prime examples of increasing river-borne nutrients and resulting diminution of coastal water quality. In 2001, more than 20,000km² of coastal waters had such low oxygen levels that shrimp and demersal fish could not survive. According to the FAO, the livestock sector is the leading contributor to nitrogen water pollution in the United States, and the FAO is convinced of the close relationship between high nitrogen levels and hypoxia (loss of oxygen). The Mississippi drainage basin contains almost all of the US feed production and industrial livestock production (Steinfeld et al, 2006). In Asia, pig and poultry operations concentrated in coastal areas of China, Vietnam and Thailand are a major source of nutrient pollution in the South China Sea.

High nitrogen pollution is a particular worry in pig producing areas, together with emissions of phosphate, potassium, drug residues, heavy metals and pathogens. The Netherlands has introduced a limit as to how

Box 4.1 *Loss of wildlife: The case of salmon*

Farmed salmon represented less than 10 per cent of the global salmon supply 20 years ago, whereas it now accounts for over 60 per cent of the market (Schouw et al, 2008). This is due both to the demise of wild stocks and to massive increases and expansion in farmed salmon production. The detrimental effect of farming on the wild population may, however, be larger than the benefit. Small-scale salmon fishermen, particularly First Nations in northern America, have long complained that waste from the farms is dangerous to wild stock, and that farmed salmon spread disease and contribute to higher concentrations of sea lice that cripple young wild salmon. After vainly communicating their concerns to the Canadian federal and British Columbian provincial governments, a First Nation lodged a class action suit against the British Columbian government. The legal action will be the first class action lawsuit in Canada launched to protect aboriginal treaty rights (Meissner, n.d.).

Farmed salmon production is largely based on a few breeding strains, developed for maximum return and fast growth. Their wild relatives, in contrast, are the products of thousands of years of evolution, adapted to challenging environmental pressures. The morphology of farmed salmon, altered for both producer and consumer, is therefore at a disadvantage in the wild. Hampered by reduced body streamlining, shorter fins, higher fat content, reduced swimming performance and differently shaped hearts, farmed salmon are less able to survive in open waters.

Escaped salmon often interbreed with wild fish stocks. This has a detrimental effect on the wild salmon gene pool, consequently lowering their survival rate. A ten-year study estimated that 70 per cent of these hybrid fish died in the first few weeks of life, whilst those that did survive could then continue to 'contaminate' the gene pool. In spite of this, there is evidence that the farmed salmon actually have a competitive edge over their wild relatives at an early stage in their lives because of their more aggressive nature and faster growth patterns. Farmed salmon and hybrids can be expected to interact and compete directly with wild salmon for food, habitat and territories, putting extra pressure on their depleting stocks. There are also threats from the transmission of diseases and parasites (Thorstad et al, 2008).

many pigs may be fattened in the country. The pig fattening industry in Europe, however, wants to grow in spite of satisfied European demand. Thus, Dutch farmers are planning new operations in Eastern Europe, particularly Poland. In Germany, there are plans to increase the number of fattening pigs by 7 per cent (Benning, 2009). Local inhabitants are up in protest, joining hands with their Polish counterparts, but will they be listened to? When will other countries also limit their pig fattening numbers?

Industrial livestock is unlikely to tackle poaching

Similarly, with regard to the exploitation of terrestrial wildlife, it has been incorrectly argued that industrial animal products could relieve the pressure on ecosystems. In its contested landmark publication, 'Livestock's Long Shadow', the FAO suggests that African bushmeat hunting could be reduced in this way. This logic, however, is full of holes. The FAO admits that bushmeat is an expensive luxury food item for the rich, while the livestock industry intends to provide low-priced food. What has made bushmeat consumption come down so far is the consumers' fear of wildlife diseases like Ebola and severe acute respiratory system (SARS), not the availability of industrial meat. It has been suggested that poached bushmeat could, however, be replaced by ranching the animals currently poached for bushmeat (Steinfeld et al, 2006). But the animal welfare implications of this are massive.

The value of local breeds

A good education for a child in an industrialized country may include a visit to a farm where colourful flocks of poultry range freely and teach their chicks what to eat, intelligent-looking pigs enjoy rooting in a field and rare breeds of ruminants forage in pastures. Seeing a work horse following the instructions of its owner to pull logs out of the forest is another breathtaking experience. Local breeds are usually well adapted to their owners and to the local environment. Many have adapted to the harsh conditions of extreme environments and have learnt to cope with heat, water shortages, fodder scarcity and diseases. In Africa, for example, one of the most economically damaging livestock diseases is trypanosomiasis. While modern cattle have no resistance to it, at least 16 African cattle breeds, including the N'dama, tolerate it.

In Rajasthan, the Raika keep sheep breeds that can cope with extreme climate. They keep several breeds at a time, in order to reduce the risks in the harsh environment of the Thar Desert. Along with a hardy breed that ensures a minimum income in bad years, they keep a higher-yielding local breed to increase their income in favourable years. The productivity of the herd is more important than the productivity of individual animals. Other criteria for selection are good maternal behaviour, walking ability and manageability. These carefully chosen breeds are the result of many generations of breeding work (Geerlings, 2001).

Local breeds fulfil many of the needs of smallholder communities and are not raised exclusively for yield. For example, draught animals are the most important source of energy by providing manure and rural transport. Some 300 million draught animals are used in Asia alone. Around half of all cultivated land in developing countries (not including China) is farmed with draught animals. Other uses and benefits include food, fibre, fertilizer, fuel, cash income, savings accounts, employment and the use of common property. Livestock keeping allows access to community-owned land. Many social functions such as dowries are tied to the herds and it is often women who use and develop breeds. The woman's role in the family changes dramatically when local breeds disappear. To be recognized, these features require a focus on people, rather than animal productivity, and they have therefore largely remained invisible.

A breed is not defined by its gene sequence nor even by its appearance – although a combination of visual features is used to describe it – but rather by the communities who have bred it. They may be nomadic or transhumant (partly sedentary) pastoralists, or farmer communities. Over the course of the last century, cooperatives and governments have engaged in breeding. They developed specific lines from existing high-yielding breeds and used new technologies like artificial insemination and targeted selective breeding for productivity. Private companies emerged a few decades ago and became globally active. They turned to selling brands, not breeding lines. Biotechnologies were developed for livestock. Poultry breeding companies concentrated down to a handful. In the last few years, livestock 'genetics' companies replaced 'breeding' companies. These combine several businesses like cattle and pig, poultry and pig, or poultry and fish. The question of who owns the breeds, or rather the genes, has become a contentious issue since these companies have started claiming patents and other intellectual property rights.

Over the course of 10,000 years of agricultural history, 40 terrestrial animal species were domesticated and many thousands of breeds developed. First sheep and goats were bred in the Near East, followed by taurine (humpless) cattle 9000 years ago. The oldest humped Zebu cattle was

Box 4.2 *Counting the breed loss*

Despite their enormous potential contribution to sustainable development and to reducing hunger and poverty, animal genetic resources for food and agriculture are underutilized and under-conserved. Of the 7600 breeds reported to the FAO by its member countries, more than 1500 are at risk of extinction or are already extinct. During the first six years of this century, more than 60 livestock breeds – almost one a month – disappeared forever, taking with them their unique genetic make-up. Losing these breeds is like losing a global insurance policy against future threats to food security. It undermines the capacity of livestock populations to adapt to environmental changes, emerging diseases or changing consumer demands (FAO, n.d.).

archaeologically evidenced in Pakistan. The pig was domesticated at several locations, including the Near East and the Far East. Asia is also the cradle of most poultry species, including chicken, duck and others. The Bactrian, or two-humped, camel, the yak and the buffalo were also domesticated in Asia. The first fish farms with various carp species were evidenced in China and date from around 4000 years ago.

Over the last century, around 1000 breeds (15 per cent of documented breeds) have disappeared, first in industrialized and later also in developing countries, accelerating the loss. What happened?

Breeding in the west turned to maximizing productivity traits: how many eggs or litres of milk an animal could produce, how much weight broilers, pigs or beef cattle could gain per day. Farming began providing the narrow optimum conditions under which breeding lines would perform based on selection from a few males. These animals were even exported to developing countries, often by development projects. It was considered too time-consuming to implement selection programmes for local breeds. Cross-breeding was thought to bring the necessary local adaptations to foreign livestock. Today it has become evident that this was not a good idea. The animals could not produce as expected, as conditions were seldom optimal: different temperatures, altitudes, humidity and different pests and diseases often decimated the foreign animals. Because local breeds were so derided, many of them disappeared before their true value was recognized by scientists and policy makers.

Indeed, western science was not alone in working against local breeds, as policy makers followed the advice of scientists. In many developing countries, factory farms and feed mills now enjoy tax breaks and subsidies. In Vietnam, 15 different subsidies are available to establish modern pig farms, totalling US$31 per sow per year. They provide 19–70 per cent of the gross margin, comparable to the subsidies available in Organisation for Economic Co-operation and Development (OECD) countries (Drucker et al, 2006). China's government appears to favour the replacement of smallholder pig production by factory farms within the next five years, and provided subsidies in 2008 to the order of 2.5 billion Yuan (US$350 million). Provincial governments have also invested in projects such as a 500,000 sow farm – one of the world's largest – in Hubei province (Xinhua, 2007).

New regulations with regard to grazing grounds have severely affected the numbers of grazing animals. For example, livestock was banned from the Aravalli hills when this region in the northwest of India became a nature park. Livestock keepers who had sent their camels there for summer grazing since time immemorial had to slaughter or sell off, and the number of camels dropped by half. All over the world, pastoralists were pressured and accused of 'overgrazing'. Nowadays, the International Union for Conservation of

Nature (IUCN) is one of the strongest supporters of pastoralism. Ecologists have established that herding improves biodiversity of grasslands and that grasslands have evolved together with herds of ruminants. Other scientists have found that pastoral communities have their own regulations to conserve the resources they live on, and that these regulations worked well until outsiders who had the power to change the rules decided they knew better.

Animal health regulations against local breeds

Since the advent of industrial breeds, traditional breeds have been blamed for spreading diseases. In fact, local breeds have always carried pests and diseases, but have used their immune systems and vitality to withstand them. In contrast, industrial livestock breeds have lost much of their resistance and vitality due to selection for high meat, milk or egg output. Their immune system remains untrained because in 'bio-secure' factory farms technical and organizational precautions are taken to prevent infections. Exposure to pathogens thus becomes a problem. Moreover, animal diseases develop in the crowded housing conditions found in factory farms.

In The Netherlands alone, more than 20 million animals were culled to prevent the spread of avian influenza, 'bird flu', including hobby birds and rare breeds. Bird flu was widely used as an apparent pretext to eradicate backyard poultry. In March 2007, Birdlife International clearly established that 'bird 'flu follows trade, not migration routes'. Only after hundreds of thousands of local poultry had been culled, the FAO concluded from a 2006 study that 'against expectations, backyard flocks in Thailand show the lowest risk of detected infection with the virus, only one quarter that of layer and broiler flocks'. Nevertheless, a year later, Gordon Butland, on behalf of the world market leading poultry breeder Aviagen/Erich Wesjohann, still claimed that 'the problem comes from backyard production, in Thailand and elsewhere' (PoultryClub, n.d.).

Brazil has become the world's leading exporter of industrial livestock products; one of the reasons for this is the country's vast feed grain producing potential. Large subsidies, tax exemptions and favourable regulations paved the way for industrial livestock keeping; smallholder livestock keeping was outcompeted. Animal health regulations were an important factor in bringing this about. The rules of the World Organization for Animal Health (OIE, Office International des Epizooties) strongly support preventive culling. There is certainly nothing wrong with setting global rules and applying them. The problem is in the rules themselves. According to the regulations of the OIE, exported animals and products must be free of notifiable diseases without vaccination, since it is not possible to tell which animals are infected and which are vaccinated.

This entails preventive culling whenever diseases occur, in order to maintain the export status. Smallholders have lost out in Brazil, have not benefited, and are unlikely to benefit from export-oriented policies elsewhere. In the current phase of global trade expansion, several other countries like Vietnam have large programmes aimed at such exports, and many others facilitate the establishment of industrial livestock production with a view to possible exports. However, large importing countries like Russia are fast stepping up their own production, and consumption in the developing world cannot be expected to reach current western levels. Furthermore, preventive culling has devastating effects, not only on agricultural biological diversity, but also when it comes to poverty alleviation and food sovereignty.

In 2007, 180 FAO member countries negotiated a Global Action Plan with a view to conserving local breeds. They all agree that these breeds are important in maintaining the viability of our food systems. In particular, climate change pressures might favour the use of traditional breeds, which are generally more resistant or tolerant to diseases and more resilient to temperature changes (FAO, n.d.). FAO member countries have agreed 'where possible', 'when appropriate' and 'if required' to establish gene banks to support on-farm conservation and to upgrade knowledge and institutions. However, industrial livestock production is not mentioned in the plan, although it was identified as a major factor responsible for breed loss in the State of the World's FAO report, 'Animal Genetic Resources' (FAO, 2007b).

The farmers quoted above joined FAO discussions at the fringes of official negotiations. Coming from the west and the developing world respectively, they voice two aspects of an approach that all smallholder livestock keepers share. Six hundred and forty million smallholders and 190 million pastoralists raise livestock worldwide, according to the FAO. Seventy per cent of the world's poor are livestock keepers. Farmer and civil society

Box 4.3 *Farmers' voices*

The industrial model of production is not sustainable. We cannot keep importing genetically modified soya beans from Brazil in order to feed poultry in the EU, which is then dumped on third markets in the developing world – forcing Brazilian farmers to over-exploit their land, EU farmers to pollute land near factory farms, and driving small-scale developing world farmers out of production. (François Dufour of Confédération Paysanne, a French farmers' organization)

We are always being told that our animals are not productive ... but we believe that an animal needs above all to be adapted to its environment. (Bouréima Dodo of the Association for the Re-dynamization of Livestock in Niger (AREN)

organizations, scientific as well as development bodies point out that it is not big factory farms and multinational corporations but small-scale family farms that hold the key to a more sustainable production process and more employment. They were backed by the 400 scientists who compiled the International Assessment of Agricultural Science and Technology Development (IAASTD, 2007).

Why gene banks may be a problem, not a solution

Conserving rare breeds in gene banks is part of the Global Action Plan agreed by FAO member countries. A lot of research money is currently being invested to make this strategy work. Technology enabling the storage of semen has existed for decades; storing eggs and embryos is far more difficult. Storing DNA – the proteins that make up genes – is another approach, popular at a time when scientists look for biotechnological solutions to the particularly burning problem of livestock disease. A rare breed may carry a useful gene that could be transferred into an industrial breeding line. However, disease resistance is not usually coded by a single gene. Older breeds often tolerate rather than resist diseases due to their strong immune systems, which are not genetically transferable. Immune systems grow with exposure to diseases. This is also the reason why rare breed tissue stored in gene banks may become useless within a short time. Frozen tissue cannot adapt to ever-changing diseases.

Who owns genes? The expansion of gene banks has added to the problem created by geneticists and companies with their patent applications. Patents, originally intended to support inventors and make their inventions available, have now led to stagnation in the area of biotechnology. A minefield of increasingly broad patents exists and researchers have to spend time and money trying to avoid trespassing. Patents on life were first granted for micro-organisms as well as biotechnological methods and genetically modified (GM) materials. Nowadays, patents are also applied for when it comes to non-GM plants and animals. Gene banks provide materials and genetic information that may hasten the process.

But the biggest problem with gene banks may be that they create the erroneous impression that breeds are being safely conserved. Animal gene banks have attracted funding that is not being spent on a far better form of conservation: directly on farms.

Could local breeds make a comeback?

In the developing world local breeds are still present. For example, 34 per cent of the chickens in Thailand are native, and in the Philippines as much as 60 per cent. But the coexistence of smallholders with factory farms seems destined to remain temporary, as several examples in the domain of pig and poultry farming have shown (Gura, 2008). It may not be possible to stop the industrialization of Chinese pig farming, mentioned above, nor the Pakistan dairy plan, which envisages the world's fourth largest milk-producing country replacing its buffaloes by industrial breeding lines of dairy cows. Nestlé, which is involved in the plan, is building the world's largest dairy factory in Kabirwala, Pakistan. Industrial livestock production is growing seven times faster than smallholder production, according to the FAO. The reasons for this are probably to be found in unfair health and other regulations as well as the massive subsidies. It is unlikely that any of these will be given up in the near future.

In the west, however, awareness of the need to move away from factory farms is growing. In Germany, two traditional pig breeds, Buntes Bentheimer Schwein and Schwaebisch-Haellisches Schwein, have been multiplied and are being raised by several dozen farmers to cater to growing niche markets. Consumers are prepared to pay higher prices and to eat less but higher quality meat. Consequently, they don't have a problem with the high fat content of the pork. Were environmental damage factored into the price of eggs, milk and meat, organic products would fare better. A German study showed that:

- Conventional pork production is subsidized in Germany to the tune of billions of Euros per year.
- External costs of conventional pork are 0.34 to 0.47 €/kg higher than organic.
- The consumer price difference between organic and conventional pork stems largely from the distribution and processing costs – economies of scale that would be reversed if preferences shifted to organic meat (Korbun et al, 2004).

Organic poultry farmers currently use the same breeding lines as conventional farmers. A large enough production of alternative poultry breeds is not available. In Germany, the public is upset by the fact that all the male chicks of layer lines are culled by the thousands every day.

The last argument that proponents of industrial livestock production resort to, apart from that of increasing demand from an increasing global population, is that young rural dwellers don't want to carry on with the hard life of their ancestors. But, if they are asked, many object to the lack of a level playing field. They can't compete with adverse subsidies, animal health regulations, grazing restrictions or settlement policies and therefore find it hard to see a future in rural areas and local breeds.

Corporate market penetration and domination

The industrialization of livestock production has reached most countries in the world. Three quarters of the world's chicken, two thirds of all milk, half of the eggs and one third of the pigs are produced from industrial breeding lines, in other words, genetically very similar animals bred for high output. Most of this production uses concentrate feed with frequent chemical veterinary treatments and often takes place on large, climate-controlled farms with increasingly heavy 'biosecurity' – measures controlling entrants to factory farms such as personnel, visitors, feed and replacement animals – to prevent infections.

Over the last decades cows, pigs and chicken bred for high levels of productivity, have been introduced into developing countries, often aided by development cooperation and supported by measures such as subsidies, veterinary services, local research and animal health regulations. Where environmental conditions were too harsh for the foreign animals to produce or even survive, cross-breeding with local breeds was advocated, as local breeds and production systems were usually considered too backward and inefficient by themselves.

Poultry and pig factories integrated into corporate value chains are growing fast in Asia and Latin America and poultry factories are sprouting up in many African countries. The four globally active poultry breeders – Erich Wesjohann Group (D), Hendrix Genetics (NL), Groupe Grimaud (F) and Tyson (US) – have established multiplication and distribution systems for their hybrid lines in all these areas. Farmers cannot breed the hybrid lines, but need replacements for each production cycle and this dependency – often contractually exclusive – has fostered an extreme concentration of business in the hands of only four multinationals. With the help of hybrid pig lines there is a rapid concentration taking place in the pig breeding industry, which is also spreading its – often exclusive – multiplication and distribution systems worldwide. The high achievable rates of return have motivated seed corporations like Monsanto to invest in livestock genetics, including cattle and pigs. Forward contracts and exclusive access to gene and information technologies, as well as patents, are also fostering further concentration.

Box 4.4 *Formation of multi-species livestock genetics corporations 2005–2008*

- 2005: Genus plc set up to combine global market leaders in pig and cattle genetics

- 2007: Hendrix Genetics (layer, broiler) buys Nutreco's breeding section (broiler, turkey, pig); in 2008, Hendrix Genetics buys France Hybrides (pig)

- 2008: EW Group (layer, broiler, turkey) acquires leading salmon/trout breeder AquaGen

- 2008: Groupe Grimaud (poultry) sets up Pig Genetics Development Company and buys shares in Newsham's (former Monsanto's) pig business.

Source: Gura (2009b)

Corporate market power exceptionally high in livestock genetics

The poultry genetics industry condensed down from a dozen globally active corporations to four within a few years. In recent years, multi-species livestock genetics corporations formed and, though medium-sized in terms of turnover, their market power is exceptional. EW Group provides 68 per cent of world demand for white egg layer genetics; Hendrix Genetics caters for around 60 per cent of brown egg layer genetics. The global broiler genetics market is shared by the same two companies plus Groupe Grimaud and Tyson. The turkey genetics market is catered for by EW Group and Hendrix Genetics, as well as a third company, Willmar. Pig and cattle genetics market leaders, both belonging to the same corporation, Genus plc, already hold large market shares. Large countries like China and Russia are developing their factory farms at high speed, using the same corporate genetics. The process is also fostered by the privatization of public breeding organizations that were common in industrialized countries and still exist in a number of developing countries.

The market power of food corporations remains a major problem, which has not yet been properly addressed by mainstream organizations. While one of the most influential development policy documents, the World Development Report 2008, was critical of the role of transnational corporations in developing country agriculture and in particular the problem of their excessive market power and resulting market distortions, it kept silent on the crucial question of how to regulate that power. No independent multilateral antitrust body exists that could scrutinize mergers and acquisitions and prevent food corporations from abusing their market power.

Box 4.5 *Market penetration of industrial breeding lines*

Currently, one-third of pigs, half of the global laying hen flock, two-thirds of dairy cows and three-quarters of broiler chickens are produced using industrial breeding lines.

Source: FAO (2007b)

In the near future, these proportions are likely to increase rapidly. For example:

• Chinese government policy appears to support the replacement of smallholder pig farms by factory farms within 5–10 years. This could mean half of all pigs globally will be produced using industrial breeding lines;

• In the fourth largest milk producing country, Pakistan, water buffaloes are to be replaced by industrial dairy cows. Nestlé is building the world's largest dairy factory in Kabirwala, Pakistan.

Source: Gura (2008)

Genetic uniformity already very high

The main industrial breeds of cattle, pigs and poultry have been reduced to a very narrow genetic pool. Although millions of animals of Holstein, Jersey and other dairy breeds exist, their genetics correspond to less than 100 animals. Population geneticists consider rare breeds with such a small gene pool to be endangered. In the US it was found that major industrial pig breeds like Pietrain, Duroc or Hampshire have similarly narrow gene pools (Blackburn et al, 2005). In poultry, the breeding lines are kept within a few corporations as trade secrets, and independent information on their genetic diversity is unavailable.

Technology to freeze and transport cattle semen has been available for more than 50 years. Moreover, with a bull's lifetime semen supply, enough cows can be artificially inseminated to produce around a million offspring. Farmers are free to choose their preferred bull until his semen sells out, and selected offspring of the same bull is used to inseminate more cows. Globally, only 2000–3000 bulls per year are evaluated and some of them eventually become semen suppliers. Selection focuses on very few traits, rendering the gene pool extremely narrow: production increases are the main objective. Increasing the number of litres of milk per day and boosting its fat content is the overriding concern in dairy cattle breeding. Fast growth and a high proportion of muscle flesh are the features sought after in beef cattle.

Selection for the same traits has taken place for many generations. Other important traits like vitality, fertility and mothering behaviour have been largely lost. On the way to enhanced productivity, the cattle developed productivity-related problems. Their overused udders producing 10,000 litres of milk per year get infected easily; their legs and hoofs are easily deformed from standing in sheds instead of walking on pastures. After two to five milking periods (cows start lactation after a calf is born) they have to be slaughtered. The normal life span of cattle is 20 years; in the US, it has come down to less than five.

Pig semen is less easily preserved than cattle semen. Artificial insemination of pigs was developed only more recently and frozen pig semen is not very effective. The reason why boars now increasingly inseminate containers rather than sows is twofold: breeding companies wish to control reproduction by selling semen, not boars, and disease vectors are better controlled in semen than in live animals. Selection in pigs is for meat leanness and uniformity, feed efficiency, daily weight gain, litter size and mothering ability to reduce the deaths of piglets before weaning. The number of weaned piglets per sow per year has risen from 25 to almost 30 in some breeding lines. 'For the past 5 years, annual genetic improvement was $3.90 per slaughtered pig' (Working Group FABRE Technology Platform, 2006).

Nowadays pig production is split into three businesses: the breeders, the multipliers and the fatteners (the 'breeding pyramid'). Specialization, as well as contracts between these sectors, also increases concentration in the pig breeding business and decreases genetic diversity. Contracts bind customers to the supplier and further reduce competition (Gura, 2008).

In poultry, genetic uniformity may be extremely high. Only four globally operating breeding companies remain after about 15 years of consolidation. They market separate breeding lines for broiler and layer hens – only about 30 different chicken breeding lines in total. Information on their genetic holdings is a trade secret and independent information is unavailable.

Specialization in poultry took place earlier than in pigs. As well as a separation of egg and meat production, breeders keep the great-grandparent and grandparent stock and send the parent stock to multipliers, providing only male chickens of the male line and female chickens of the female line to exclude the possibility of breeding by the multipliers. The multipliers then cross the parents and send the eggs to hatcheries, while the hatcheries supply one-day-old chickens to egg and broiler producers. It is estimated that, every day, millions of one-day-old chicks travel around the world in cardboard boxes. This is possible because they don't need feed or water for some time after hatching.

Hybrid technology has added to the loss of diversity. The great-grandparent and grandparent stock kept by the breeding company comes from

Box 4.6 *Hybrid livestock – a tool for market development and domination*

Hybrid chickens were first developed in the 1940s by Henry A. Wallace, who was the 33rd Vice President of the United States (1941–1945). Henry Wallace applied the same breeding methods to poultry that he had used to develop Pioneer Hi-bred corn. When two different breeding lines are cross-bred, productivity of the offspring can increase due to hybrid vigour. However, this 'heterosis' effect gets lost in the next generation, so that farmers using these breeds have to buy new breeding stock every time. It took only ten years for all commercial poultry breeders to breed poultry hybrids. Hybridization allows not only higher productivity, but also more market control if the original lines are kept exclusively in the breeding company. Now, hybridization has become common in pigs and is occasionally applied in aquaculture as well.

specialized inbred breeding lines. One 'product', usually branded – a broiler or an egg layer – is genetically made up by four inbred lines. The chickens are selected for male broiler features such as large breasts and high daily weight gain, and female features such as large egg numbers and reaching egg-laying age early. Their offspring (parent stock) are sent to multipliers for cross-breeding and the next generation is particularly productive. One great-grandparent rooster can have 28 million offspring in one producing generation and an industry insider claimed that the number of pedigree animals needed to supply the world with chicks may fit into a garage.

Modern broilers grow three times faster than 30 years ago and require less time and feed to reach slaughtering weight. Egg numbers have increased from less than 270 to 340 per year within the second half of the last century. Feed efficiency – the amount of produce per kg of feed – has increased by one third. But because of genetic selection, skeletal problems occur frequently and the animals are susceptible to stress and diseases. They struggle to survive in conditions other than the optimum altitude, humidity and temperature. And, as in pigs, their immune system is compromised to such an extent that they often need antibiotics. Hybrid turkeys must also be artificially insemi-nated, as males are impaired by their excessively heavy breasts and cannot mate with female birds naturally. Expenses associated with biosecurity to protect weak animals from infections are constantly increasing. Epidemics are becoming increasingly costly, and the EU is currently negotiating an agreement meant to shift their cost from the farmer to the taxpayer. In Asia, the economic losses caused by avian influenza are estimated at US$10 to 15 billion. By 2005, 140 million animals had been culled.

Box 4.7 *New reproduction and selection technologies*

New reproduction and selection technologies are leading to:

- higher selection intensity (e.g. DNA marker assisted selection);

- shorter generation intervals (e.g. embryo selection);

- more females than males in cattle and pigs ('sexed semen');

- replication of the same (clones), in other words, a faster increase in genetic uniformity.

Source: Gura (2009b)

New technologies are leading to higher genetic uniformity

New reproduction and selection technologies that are expected to further reduce the variation within industrial breeding lines are now available. Marker assisted selection will increase selection intensity. Selecting from embryos rather than waiting for the animals to reach adult age will shorten generation intervals. 'Sexed semen' is now available that helps farmers to generate fewer male animals that are considered useless for production, but will also further reduce the variation in cattle or pigs. Cloning aims to replicate the same genetic set-up and eliminate what is left of natural variation.

The genetic modification of animals has made enormous regulatory and technical progress in recent years. Genetically engineered chickens appeared in the 1980s and experiments continue. Particularly interesting to livestock genetic companies is genetic engineering for disease resistance, and to improve behaviour in cages, a major animal welfare problem.

Many GM creations, such as a pig with human growth genes, or goats who produce spiders' silk in their milk, have failed. Both of these had received US public funding, and the failures may never be accounted for publicly. Pigs' saliva and digestive tracts have been altered by genetic engineering to reduce the phosphorus in pig waste. In order to make pigs produce more omega-3 fatty acids relative to the less healthy ones, a roundworm (*C. elegans*) gene was transferred into the pig genome. To increase milk production and the piglets' ability to digest milk, a cow gene and a synthetic gene were introduced into the pig genome. Jersey dairy cows received a staphylococcus bacterium gene to fight the kind of bacteria that often infects the overstrained udder. The cows' own milk production genes were tripled or quadrupled to make them produce ever more milk (information made available by Jaydee Hanson, Center for Food Safety, pers comm, 2009).

Consumers accept genetic engineering for medical purposes more readily than for food. 'Pharm animals' such as cattle and goats that secrete human antibodies and growth hormones in their milk were the first terrestrial animals to receive market approval. A genetically modified animal, a goat that excretes a blood thinning drug in its milk, was approved for the first time in Europe in May 2008 and in the US in February 2009.

The legal situation regarding genetically modified animals is not specifically regulated in the US. The animals are simply tested to see whether they tolerate the inserted gene and whether the GM products are safe to eat. No labelling, no import regulation, no tracking is required. Market applications are kept secret by the Food and Drug Administration (FDA) until approval. European Union regulations stipulate that neither animal health nor the environment should be affected, and require labelling, among other constraints (see box below).

Consumers in the US may thus eat food of GM animal origin unknowingly. But some may not want to eat GM animal food – not even if it's free. The biotech company Pharming Healthcare in Philadelphia suggested donating the animals' meat to a local food bank, but local area residents questioned its safety and the company was forced to withdraw its offer (information made available by Jaydee Hanson, Center for Food Safety).

A GM ornamental fish received market approval in the US, but not in Europe. Approval of GM animals for human consumption has been denied for the last ten years, when market approval for a fast-growing GM salmon was applied for in the US. The biology of reproduction is much simpler in fish than in mammals, and gene transfer methods in fish are therefore significantly more advanced. Around 35 aquatic species, including carp, salmon, trout, tilapia, shrimp, oyster and abalone have so far been genetically modified, usually in order to hasten and increase growth, with the involvement of research organizations and companies in the US, Canada, New Zealand, China, India, Korea, Cuba and Belgium (Mair, 2007). These experiments give a foretaste of what may become the norm commercially in the future.

GM animals: Risks to human health, animal welfare and the environment

A number of risks to human health have been pointed out:

- New allergens, hormones or toxins can be created when genes are inserted into the cells of animals.

Box 4.8 *The legal situation of GM animals*

In the European Union EU Directive 1829/2003 requires that:

- no negative impact on human or animal health or the environment occurs;

- consumers are not misled; and

- the product is no different from equivalent food so that consumers will not suffer from deficiencies if usual amounts are consumed.

EU Directive 1830/2003 requires that genetically modified food is labelled.

In the United States:

- There are no direct regulations, only non-binding guidance of January 2009 (a few days before the Obama administration took over);

- Genetically modified animals are viewed in the same way as a 'drug': 'New Animal Drug' regulation applies but tests only to see if the inserted gene does not cause health problems for the genetically engineered animal;

- Transparency regulations are not clarified: FDA policy is to hold public advisory committee meetings prior to making decisions, but the application has to be kept secret until the 'drug' is approved;

- There is no special labelling: FDA maintains that GM foods are no different from conventional foods and thus there is no need for labelling. A label is only recommended if the food of GM animal origin has new nutritional content, but the label will only provide information on the new properties and will not necessarily state that the product is genetically engineered;

- No tracking is required of the GM animal.

- The spread of animal viruses to humans – HIV, chicken and pig flu viruses, prions that transmit 'mad cow' disease to humans; biotech companies have no means of identifying or eliminating pathogens that might spread to humans.
- New diseases in livestock and humans can be created if viral vectors, used to invade the cells and deposit new genetic materials, recombine with a virus in the animal (information made available by Jaydee Hanson, Center for Food Safety).

Animal welfare is at risk because:

- EU and US regulations on genetically modified organisms do not consider animal welfare.

- GM animals may exacerbate the inherent problems of factory farming and introduce new potential hazards.
- In the process of DNA microinjection, there is no control over where the genes go; errors can cause deformities and other genetic defects.
- Less than 4 per cent of GM animals survive.
- Many GM animals do not express the genes properly and have physical or behavioural abnormalities.
- Reproductive technologies (in vitro culture, semen collection, egg collection, cloning) can cause stress in animals; e.g. in vitro culture methods and cloning have been associated with 'large offspring syndrome' in cows. GM animals are often reproduced with such technologies (information made available by Jaydee Hanson, Center for Food Safety).

Finally, the environment is at risk because:

- There is a high risk of aquaculture diseases spreading to wild populations, e.g. the 2009 outbreak of salmon lice and a salmon virus in Chile.
- Escaped transgenic animals could introduce new genes into wild populations, lowering their fitness.

The latter situation could lead to the extinction of both the escaped transgenic population and of its wild relatives. For example, escaped male GM fish designed to grow larger than normal would enjoy outstanding mating success, because female fish see large males as 'fitter' than small males and the transgene would rapidly spread through the wild population. If just 60 genetically altered fish were released into a population of 60,000 wild fish, complete extinction of the wild population would follow within 40 generations (Muir and Howard, 2002). Indeed, millions of farmed salmon escape every year and the wild population is almost extinct.

These effects are not specific to GM fish, but GM fish or any aquatic or even just mobile species may exacerbate the problems. The aquatic genetics industry is offering reproduction control technologies to overcome this environmental problem. The sterilization technologies used consist of heating fish eggs or rendering the animals triploid (modifying them to possess three sets of chromosomes instead of two). This technology has been developed for many species, but is currently only being applied in trout. Aquatic genetic companies and scientists always argue that triploidy could solve the environmental problem of GM fish, the Trojan gene introduction that could kill off wild populations. The sterilization process is not reliable, however – but it does seem good enough for the genetics companies.

The Canadian government is currently funding research at the US company AquaBounty to develop triploidy for its GM salmon, which has been hoping to receive US market approval for a decade now. Triploidy has so far failed in salmon.

Why the livestock genetics industry is crazy about clones

The livestock genetics industry is particularly interested in cloning in combination with genetic modification. If a genetically modified animal is reproduced, the offspring is not guaranteed to carry the transferred gene. With cloning, the guarantee is there. Although cloning has had meagre technical success, the US administration is supporting the development and marketing of the technology and the EU appears to be moving in the same direction.

The cloning process, theoretically aimed at recreating an exact copy of an animal, is, ironically, a matter of coincidence in practice. Following 25 years of research and 12 years after the birth of Dolly, the results of cloning are consistently incalculable and unreplicable. Moreover, they are a disaster as far as animal welfare is concerned. Many cloned animals die or suffer from deformities due to unintended changes in gene regulation. The concept of life as a conglomerate of parts that can be replicated or exchanged is simply wrong. It took millions of research Euros, dollars and yen to reach this conclusion, which any taxpayer with an average knowledge of biology could have foretold, and which has not yet been discussed by applied science, according to the livestock scientist and technology analyst Christoph Then (Then, 2009).

The US Food and Drug Administration bases its policy on the theory, not on the practice. In January 2007, it ruled that cloned animals are no different from conventional animals. The FDA approved the sale of meat and milk from cloned animals, without labelling, similarly to GM animals. Even the rather conservative US National Academy of Sciences has cautioned the Federal Government to monitor cloned animals and 20 US companies have pledged not to use them. Cloning companies reacted by proposing a radio frequency tagging system; the monitoring would stop at the slaughterhouse door, however, and only cover clones, not their offspring.

In the EU, the European parliament had for many years supported all decisions that enabled cloning: research budgets and the liberalization of breeding legislation in 2005. But, on 3 September 2008, it adopted a resolution on cloning – with 622 votes in favour and only 32 against – asking the European Commission to implement bans on cloning, on the marketing of meat and milk from cloned animals or their offspring, and on the import of semen, embryos, meat or milk from cloned animals and their offspring. The parliament recognized two important problems: first, a large proportion of the cloned animals suffer very severe health problems; second, cloning will further reduce the genetic diversity of livestock and render it more susceptible to diseases.

The EU Commission delayed the issue and again asked the European Food Safety Authority (EFSA) to look into the health and welfare of animal clones in March 2009. The EFSA was given a deadline of June 2009, just after the elections to the European parliament. In July, EU Agriculture Ministers came up with a proposal to include the offspring of clones in the Novel Foods Directive. Contrary to the parliamentary ruling, the Commission thus suggested regulating rather than banning their market admission. Cloned animals were already regulated as novel food and their offspring is to be included. Twenty-three of the 27 EU countries, however, announced that they wanted a ban.

Conclusions

The new livestock genetics technologies are likely to increase the perversity of industrial livestock production. Biodiversity loss, hunger and climate change are all part of the same problem of overconsumption, where industrial livestock keeping – as well as industrial aquaculture – plays a pivotal role.

Industrial livestock policies are invariably playing on the widespread but incorrect belief that we need animal proteins for a healthy diet. Even developing countries are subsidizing industrial livestock and aquaculture production, instead of supporting smallholders. In fact, keeping livestock provides livelihoods for most of the world's poor. Developing countries should not sacrifice their ecosystems for fish or shrimp exports to countries already oversupplied with animal proteins, but instead support smallholders to produce sustainably for local markets. But developing countries should support extensive smallholder livestock keeping rather than sacrificing ever more land for feed production. The 'climate efficiency' claim of industrial livestock is wrong for many reasons, but particularly wrong with regard to feed, since the chemical fertilizers used in crop production increase dangerous nitrous oxide emissions. Besides rainforests, industrial livestock production is likely to destroy grasslands, which are not only a resource for livelihoods based on extensive livestock keeping and an important ecosystem, but also a major carbon sink due to their root mass.

Extensive livestock keeping is now often portrayed as damaging to the climate because it is not competitive in terms of narrowly defined productivity.

The climate crisis should expose our flawed production systems, but is misused to push for a further intensification of industrial livestock production, while the real problem is simple – too many industrial cattle, pigs and poultry. The United Nations should no longer refrain from exploring the possibilities of reducing livestock production.

References

Benning, R. (2009) 'Tierschutz ist Klimaschutz. BUND-Recherchen zur Tierhaltung in Deutschland', Biomar Annual Report 2007, available at www.biomar.com/investor_relations_pdf_2008/BioMar_Annual_Report_2007.pdf, last accessed 20 May 2009

Blackburn, H., Welsh, C. and Stewart, T. (2005) 'U.S. Swine Genetic Resources and the National Animal Germplasm Program', available at www.nsif.com/Conferences/2005/pdf%5CGermplasmProgram.pdf, last accessed 20 May 2009

Drucker, A. G., Bergeron, E., Lemke, U., Thuy, L. T. and Valle Zárate, A. (2006) 'Identification and quantification of subsidies relevant to the production of local and imported pig breeds in Vietnam', *Tropical Animal Health and Production*, vol 38, pp305–322

FAO (2007a) 'State of World Fisheries and Aquaculture 2006', FAO, Rome

FAO (2007b) 'State of the World's Animal Genetic Resources', FAO, Rome

FAO (n.d.) 'Animal genetic resources: A safety net for the future', available at www.fao.org/fileadmin/templates/nr/documents/CGRFA/factsheets_animal_en.pdf, last accessed 9 February 2010

Geerlings, E. (2001) 'Sheep husbandry and ethnoveterinary knowledge of Raika sheep pastoralists in Rajasthan, India', MSc thesis, Environmental Sciences, University of Wageningen, Wageningen, The Netherlands

Greenpeace International (2009) 'Slaughtering the Amazon', available at www.greenpeace.org/raw/content/international/press/reports/slaughtering-the-amazon.pdf, last accessed 20 September 2009

Gura, S. (2008) 'Industrial livestock production and its impact on smallholders in developing countries', available at www.pastoralpeoples.org/docs/gura_ind_livestock_prod.pdf, last accessed 23 September 2009

Gura, S. (2009a) 'Supporting global expansion of aquaculture. The new strategy of the European Commission', *World Economy and Development*, vol 3, available at www.wdev.eu/wearchiv/042ae69c0b1129b01.php, last accessed 27 September 2009

Gura, S. (2009b) 'Corporate livestock farming: A threat to global food security', in *Third World Resurgence,* Third World Network, Malaysia

Halweil, B. (2008) 'Worldwatch Report 176: Farming fish for the future', Worldwatch Institute, Washington DC

IAASTD (2007) www.agassessment.org, last accessed 20 September 2009

Korbun, T., Steinfeldt, M., Kohlschütter, N., Naumann, S., Nischwitz, G., Hirschfeld, J. and Walter, S. (2004) 'Was kostet ein Schnitzel wirklich? Ökologisch-ökonomischer Vergleich der konventionellen und der ökologischen Produktion von Schweinefleisch in Deutschland', Schriftenreihe des IÖW 171/04, Berlin

Mair, G. C. (2007) 'Genetics and breeding in seed supply for inland aquaculture', in Bondad-Reantaso, M.G. (ed.) 'Assessment of freshwater fish seed resources for sustainable aquaculture', FAO Fisheries Technical Paper No. 501, Rome, FAO, pp519–547

Meissner, D. (n.d.) Press News Limited, available at www.seafoodnews.com/newsemail.asp?key=499503, last accessed 20 May 2009

Muir, W. M. and Howard, R. D. (2002) 'Environmental risk assessment of transgenic fish with implications for other diploid organisms', *Transgenic Research*, vol 11, pp101–114

Norris, A. (2008) *Review on Breeding and Reproduction of European Aquaculture Species: Atlantic Salmon (Salmo salar)*, Marine Harvest, Ireland

PoultryClub (n.d.) www.poultryclub.com/uploads/media/kiev_butland.pdf, last accessed 20 May 2009

Pynn, L. (2009) 'First Nations sue over salmon', *The Vancouver Sun*, 5 February, available at www2.canada.com/vancouversun/news/westcoastnews/story.html?id=357518d8-6d67-466c-9c9e-147bf3e2c9e0, last accessed 12 May 2009

Schouw & Co. (2008) 'Annual report', www.biomar.com/Global/global%20press%20releases/Schouw2008_UK.pdf, last accessed 16 July 2010

Steinfeld, H., Gerber, P., Wassenaar, T., Castel, V., Rosales, M. and de Haan, C. (2006) 'Livestock's long shadow: Environmental issues and options', FAO, Rome

Then, C. (2009) *Dolly ist tot: Biotechnologie am Wendepunkt*, Rotpunktverlag, Zürich

Thorstad, E. B., Fleming, I. A., McGinnity, P., Soto, D., Wennevik, V. and Whoriskey, F. (2008) 'Incidence and impacts of escaped farmed Atlantic salmon *Salmo salar* in nature', NINA Special Report 36, available at www.asf.ca/docs/uploads/impacts-escapes-2008.pdf, last accessed 20 May 2009

Van Mulekom, L. (2007) 'NGO perspectives on aquaculture, aquaculture certification, and the responsibility of commodity buyers', input to FAO Expert Workshop on Guidelines for Aquaculture Certification, London, 28–29 February 2007

Van Mulekom, L., Axelssonb, A., Batungbacalc, E. P., Baxterd, D., Siregare, R., de la Torred, I. and SEAFish for Justice (2006) 'Trade and export orientation of fisheries in Southeast Asia: Under-priced export at the expense of domestic food security and local economies', in *Ocean & Coastal Management*, vol 49, nos 9–10, pp546–561

Working Group FABRE Technology Platform (2006) 'Sustainable farm animal breeding and reproduction. A vision for 2025', available at www.fabretp.org, last accessed 20 May 2009

Xinhua (2007) 'China earmarks 15.2 bln yuan to ensure pork supplies', available at http://english.peopledaily.com.cn/90001/90776/90884/6326523.html, last accessed 12 February 2010

5

Does Organic Farming Offer a Solution?

Richard Young

I am on my first tractor, ploughing through the night as only young farmers can. It is the autumn of 1969 and I have the single-minded ambition to turn 200 acres (81 hectares) of Cotswold grassland into fields of waving barley, just like those that surround me.

Now I am unloading 60 tonnes of subsidized fertilizer – 1200 sacks, each weighing over 50 kilograms, taken off the lorries on my shoulder and piled up in the middle of a farmyard still dotted with carts and hay rakes from the horse-drawn era.

The combination of ploughing and fertilizer allows me to qualify for a farm development grant and cash in on the rise in the price of grain, from £18 a tonne in 1970, to over £100 a tonne within just a few years. In addition to nitrogen fertilizer, herbicides and fungicides, I apply large amounts of phosphate fertilizers. I buy a new combine harvester and three more tractors, then take on neighbouring land, doubling the farm's size.

The scene fades, replaced by me rising from my seat at the annual dinner of the Gloucestershire Root, Fruit and Grain Society five years later to receive a prize for the best crop of barley in the county. Someone slaps me on the back as I make my way through the tightly arranged tables. Applause is ringing in my ears; it's the closest I will ever come to winning an Oscar. Johnny Morris, the popular television personality, shakes me by the hand and gives me a silver cup.

These and other memories from my early farming life have been playing in my mind during a depressing early-morning journey in the Land Rover with two faithful old friends at the end of their working lives in a trailer behind. The exhausting ten-hour round trip takes me from the North Cotswolds, where I live, to an abattoir in southern Somerset – the nearest slaughterhouse approved by Defra (the UK government Department for the Environment, Food and Rural Affairs) to kill and incinerate cows born before 1997, under the Older Cow Disposal Scheme. I have to take them alive because Defra has (in my view, sadistically) closed the loophole that once allowed us to have them shot on the farm under veterinary supervision and transported to the incinerator dead. The experience is additionally galling because the BSE crisis drove our finances into the red, even though we never used meat and bone meal (MBM) in livestock feed, have bought-in no cattle since the mid 1970s and, in common with over 70 per cent of farms with beef suckler herds, have never had a case of BSE (Defra, 2009a).

But while BSE is largely eradicated, things are still not easy for cattle farmers in the UK. Dairy farms have halved over the last decade to just 17,000 (Doward, 2009). The situation is no better for beef producers. Despite having the most suitable climate for grazing in Europe, the UK has gone from beef exporter to importing 300,000 tonnes of beef a year in little more than a decade. As the chairman of the English Beef and Lamb Executive recently said, 'The apparent lack of awareness on the part of policy-makers and the public of the steady attrition of the beef industry, means that we are sleepwalking towards the irretrievable decline of a critical part of our farming industry' (Cross, 2009).

Given the now widely recognized issue of methane (CH_4) emissions from cattle and sheep (Steinfeld et al, 2006), this may sound like good news, but is reducing the number of cattle kept in the UK really the way to a sustainable, healthy and low-carbon food system?

I am still thinking about Johnny Morris though, and the speech he gave before awarding the prizes. This was just after the 1973 oil price shock and the first rumblings by environmentalists of serious harm being done to the planet.

He talked about the rape of the earth by modern agriculture, the 'dust bowl' of 1930s America and the death of birds from agricultural pesticides described in Rachel Carson's *Silent Spring* (1962). He likened the Earth to an orange. Holding one up in his hand, he said that we lived on its very thin skin of soil, which we were denuding, had sucked out most of its juice and were busy spitting out the pips!

When I set about bringing one of the last unimproved Cotswold farms into the modern world I thought nothing about the acres of wild flowers and rare grasses I was ploughing up, or the loss of wildlife habitat this involved, let alone any possible impact I might have on the wider environment. Life was simple then. My job was to help my neighbours produce enough barley to supply the growing number of factory livestock farms, and with any grain left over we would fill Europe's intervention stores to the point of over-flowing. Unlike today, we all felt wanted and relished the challenge.

Johnny Morris's comments, however, would have a life-changing effect on me. They gave me a different view of my role as a farmer and opened my eyes to the many negative changes taking place in agriculture. A chance meeting with an organic farmer shortly after showed me, to my surprise, that there was another way to farm – one I'd never even heard of before.

Farming without fertilizers and sprays, though, wasn't easy, and if I'd grown up in the era before they became available I might not have abandoned them so abruptly. Yields plummeted, weeds abounded. My wheat no longer had enough protein for bread making and I got a lower, not higher, price for my organic barley because the grains were smaller and contaminated with weed seeds. I quickly saw my healthy profits turn to a significant loss. There were no reliable sources of advice, no premium organic prices and no incentives from the government to farm that way. By forsaking nitrogen fertilizer, I even affected my entitlement to a lucrative capital grant scheme, since the farm no longer met this essential criterion for increasing productivity.

Gradually, I learned from my mistakes, old farming books and other organic farmers what was needed to produce food successfully by this very different method. The secret was to expand the small beef herd I had retained on fields too steep for ploughing.

Once I built up the cattle numbers I could have half the arable land rebuilding soil fertility under grass, to balance the half where the soil was being exploited by growing grain. After a few shaky years both yields and quality improved, and my livestock health increased so much I no longer needed to use wormers or other routine medication. As such, I was eventually well placed to benefit from growing public support for the organic approach and higher prices for my crops and animals from the mid 1980s onwards. Since then demand for organic food has increased steadily, with sales in the UK reaching £2.1 billion in 2008 (Cottingham and Leech, 2009). But with productivity per acre still the driving force, industrial farming continues to dominate global agriculture. Even now, only 4 per cent of UK and 0.65 per cent of global farmland is farmed organically (Cottingham and Leech, 2009).

Global problems

Global food production has increased in recent decades, more or less in tandem with population growth, creating the impression that there is nothing fundamentally wrong with current farming methods and that all we need is more of the same, plus a few new technologies, to feed an estimated population of 9 billion by 2050. Yet, in many parts of the world productivity is actually falling as a result of intensive methods. Desertification, coastal flooding, salination, soil erosion and degradation, depleting water aquifers and contamination are reducing the area of productive land in regions such as the Nile Delta, southern Iraq, Spain, Australia and even parts of the US. In the UK Fens we have grown crop after crop, put nothing back and all but used up that huge grow bag of fertility. In India, the intensive use of pesticides, fertilizers and irrigation, which underpinned the inappropriately named 'green revolution', increased yields and encouraged continuous arable cropping on the most productive soils in areas like the Punjab. But yields are now falling and India has again become a net grain importer. As the Indian government states:

> *With the wheat and paddy rotation during the green revolution for producing more and more food grains, the natural resources of soil and water have been degraded and depleted…Excessive use of chemical fertilizers, insecticides and pesticides, etc., have destroyed the physical structure of the soils leading to its decreased water-holding capacity.*
> (Indian Government, n.d.)

Intensive cropping also encourages new, untreatable strains of diseases – yellow rust in wheat and foliar diseases in rice, which have the potential to spread and decimate crops over vast areas. And as the climate warms, insect pests increase, while erratic floods and droughts leave vast regions critically short of food.

To prevent a future food crisis of truly global proportions, agriculturalists not only need to deal with these and many similar problems, but crucially they also need to cut agriculture's carbon footprint. Organic farming is more sustainable in that it relies less heavily on non-renewable inputs and improves soil quality rather than degrades it, but could it feed the world, make us all healthier, increase biodiversity and also reduce agriculture's contribution to climate change?

Livestock farming

In recent years there has been growing awareness of the negative aspects of livestock production for the climate, and it is clear that if we all ate less meat and dairy products we would reduce agricultural emissions of greenhouse gases (GHGs) considerably. The focus of concern has been on sheep and cattle, due to their CH_4 emissions. But cattle and sheep are at the heart of most organic farming.

Another perceived weakness is that conversion to organic methods, at least in Europe, involves a reduction in productivity per hectare. But it gets worse. According to scientists at Cranfield University, suckler beef herds, like my own, are 50 per cent less energy-efficient at producing beef than intensive dairy herds (Williams et al, 2006), and while I naively assumed that cattle and sheep naturally grazing unfertilized grassland would emit less CH_4 than cattle reared intensively indoors, or in feedlots on grain, the accepted wisdom is that they emit more (Johnson and Johnson, 1995).

Methane is the second most important GHG, 21 times more potent than carbon dioxide over 100 years. Its concentration in the atmosphere has increased from under 700 parts per billion (ppb) before the industrial revolution to over 1700ppb today, but its relatively short life (12 to 15 years) makes it an ideal target for climate change amelioration. A rapid reduction now might just help us avoid runaway climate change. This suggests that we should all farm without ruminants.

That creates a big problem for organic farmers because ruminants make commercial use of grass and clover, which are an essential part of sustainable organic rotations. Legumes like clover and lucerne (alfalfa), which only animals can digest, can take up to 200kg of nitrogen per hectare out of the atmosphere each year and add it to the soil, far more than edible legumes such as peas and beans. Nitrogen is essential for plant growth and this is the only alternative way of introducing sufficient into the soil to grow bountiful crops. While it is possible to grow legumes and simply plough them in for fertility, this means taking land out of production – difficult to justify on a large scale in a world of potential food shortages.

Just how important grazing animals are to organic food production is illustrated by a recent report commissioned by the Soil Association from Reading University (Jones and Crane, 2009). This considers what agriculture in England and Wales might be like if it went totally organic.

Three of the report's findings are compatible with the case for eating less meat and dairy products. Based on the records of 176 organic farms, the researchers estimate that there would be 70 per cent less pork and chicken and 30 per cent less milk and dairy products. Grain production would also fall by about 30 per cent, but there would still be plenty for human consumption, because we would no longer need the vast tonnages for intensively reared livestock – all factory farms would disappear. An organic Britain could also match the yields of intensive vegetable production. So far so good. But to provide the fertility to make this possible we would need to return more land to grass and clover. To make that economically viable we would have to introduce cattle and sheep on to prime arable land, increasing, not decreasing, overall beef cattle and sheep numbers – in the example of the Reading model, by 68 and 55 per cent respectively. This would turn us back into the net beef exporter we used to be and make us self-sufficient in lamb, which would, of course, increase the UK's CH_4 emissions. Another seemingly unhelpful aspect to the organic case is that while poultry emit little more CH_4 than humans they would dramatically decline in numbers.

The first detailed life cycle analysis of GHG emissions associated with organic and non-organic farming found that in terms of carbon dioxide equivalent (CO_2e), producing a kilogram of organic beef or lamb puts three to four times as much carbon into the atmosphere as producing a kilogram of chicken from an intensive indoor system (Williams et al, 2006, p72).

This research is influencing Defra's developing strategy and even caused David Kennedy, chief executive of the government's Committee on Climate Change, to give up eating lamb and encourage us all to do likewise, while increasing still further our consumption of pork and chicken (Leake, 2009).

The Reading team has also assessed what agriculture would be like if it changed to meet the government's vision of healthy eating. There would be a big increase in cropping, with an additional 100,000 hectares of grazing land ploughed up in southwestern England. As with the organic option there would be a reduction in dairying, but milk would come from even fewer, ultra-high-yielding cows. In stark contrast to the organic option though, beef and sheep would decline by 20 per cent and disappear entirely from most upland farms (Jones, 2009).

With climate change rightly becoming the principal driver of agricultural policy does this mean that organic farming, with its high dependence on grazing animals, is not fit for purpose?

Though not yet articulated, this is the natural extension of recent UK government thinking. Writing to a fellow MP, the former Defra Minister Jane Kennedy stated, 'While all livestock production is associated with GHG emissions, the more intensive and efficient livestock production systems generally have lower emissions per kilogram of meat or milk produced than the more extensive free-range type of systems' (Kennedy, 2009). Research commissioned by Defra has also claimed that greater intensification of milk production is good for the climate because higher-yielding cows fed high-energy diets produce more milk per litre of CH_4 they emit (Genesis Faraday, 2008).

It's beginning to look as if it's me who is ruining the planet, not my intensive farming neighbours. But before finally being pushed to sell our herd of beef cattle, plough up all our grassland and buy a fertilizer spreader again, I'm going to take one further look at the evidence, starting this time with a GHG that normally receives little attention in relation to farming.

Carbon dioxide – the hidden evidence

Trying to understand the overall impact of agriculture on the climate has been occupying scientists for decades. Attention has focused on CH_4, and to a limited extent nitrous oxide (N_2O), officially UK farming's main GHGs; but what about CO_2? It may seem like a daft question, but CO_2 is responsible for more than 80 per cent of anthropogenic global warming emissions. On first view it appears that farming accounts for only a tiny part of UK emissions, as can be seen from Table 5.1.

But this is not the full story. There are several areas where significant emissions of CO_2 and N_2O are usually ignored by commentators or obscured in agricultural inventories. The first and most obvious relates to energy use in

Table 5.1 *GHG emissions in the UK in 2007, by source, as millions of tonnes (mt) of carbon dioxide equivalent (CO_2e)*

	UK total	Contribution from agriculture		
	(mt)	mt	% UK total	% agriculture
Total CO_2 equivalent	636	48	8	
CO_2	543	4	<1	9
CH_4	49	18	3	38
N_2O	35	25	4	53
Other	9			

Source: Calculated from Defra (2009b)

agriculture: fuel, electricity, etc. Data is available from Defra, and including it increases CO_2 emissions by 2.4 million tonnes. This, though, does not include the emissions associated with the manufacture and transport of inputs such as fertilizers and pesticides, indispensable for all non-organic farming. For fertilizers made in the UK, the emissions are listed partially with industry and partially with transport figures. However, about half of all nitrogen fertilizer we use is imported. The emissions associated with this are significantly higher, due to the use of less energy-efficient factories, but they are not included in UK tables at all.

The combined CO_2 and N_2O emissions associated with the manufacture of synthetic nitrogen in the UK is equivalent to 6.7 tonnes of CO_2 for every tonne of nitrogen used in fertilizer (Mortimer et al, 2003). British farmers currently apply about 1 million tonnes of nitrogen each year in the UK. Including 50 per cent more for emissions for imported fertilizer we need to add an additional 8.5 million tonnes of CO_2e (of which 70 per cent comes from N_2O) to the UK's true agriculture GHG budget.

The significance of this can be seen from a comparison of the relative energy efficiency of organic and non-organic farming. With only limited exceptions, organic farming is significantly more energy efficient than non-organic farming (see Table 5.2).

This comparison includes fertilizers and fossil fuels, but not agricultural soils, the third major area where emissions are usually ignored, as the Cranfield researchers acknowledge in the final paragraph of their report (Williams et al, 2006, p87). Ploughing up established grassland for crop production leads to a substantial loss of carbon as CO_2 and nitrogen, in the forms of nitrate leached into water and N_2O emitted into the atmosphere. In a series of annual reports for Defra, the Centre for Ecology and Hydrology (CEH) has shown that the average carbon losses from converting grassland

Table 5.2 *Comparison of energy use in organic and non-organic production*

Crop	Energy use per tonne of food in organic v. non-organic agriculture (%)
Milling wheat	84
Potatoes	103
Carrots	75
Cabbage	28
Onions	84
Broccoli	51
Leeks	42
Beef	59
Sheep	43
Pig meat	65
Milk	72
Eggs	110
Poultry meat	111

Source: Cormack and Metcalfe (2000) for carrots, cabbage, onions, broccoli and leeks, otherwise Williams, (2007)

to continuous cropping in the UK range from 23 to 90 tonnes per hectare (84 to 330 tonnes CO_2) depending on the level of soil organic matter (CEH, 2009). Annual carbon losses steadily decline over 100–150 years, eventually stabilizing at a new lower level. N_2O is released much more rapidly than CO_2, but since it persists in the atmosphere for over 100 years its effect is just as long-lasting. Some scientists even predict that it will eventually become as significant as CO_2 for the climate (Melillo, 2009). The CEH has not yet produced data for N_2O losses from grassland conversion in the UK, but Dutch scientists have calculated that the combined effect of the CO_2 and N_2O lost to the atmosphere from ploughing long-established grassland in The Netherlands is equivalent to about 250 tonnes of CO_2 per hectare (average 2.5 tonnes of CO_2e per hectare, each year for 100 years), with N_2O accounting for about a third of this (Vellinga et al, 2004).

Since 1850 approximately 10 per cent of all the carbon added by humans to the atmosphere has been lost from soil in this way (Houghton, 1999).

As with the GHGs associated with fertilizer production, the CEH data on the loss of carbon from grassland conversion is also not included in the agricultural GHG inventories, but in yet another category; this time one related just to land use change, where it is hidden away amongst figures on carbon sequestration in forestry plantations. The data indicate that just over 6 million tonnes of CO_2 have been added to the atmosphere every year since 1990, due to the net increase in land converted to cropping. In order to include N_2O, we should tentatively increase this to 9 million tonnes of CO_2e, based on the Dutch study.

To this should also be added the smaller amounts of carbon still being lost annually from the millions of hectares in continuous cropping before 1990, as well as the N_2O that was released at the time. Reliable data is hard to find, however, except for the Fens where 3 to 4 metres of peat has been lost since 1850 and over 1 million tonnes of CO_2 (380,000 tonnes carbon) is still being emitted annually (Williams et al, 2006). Other emissions could be added relating to pesticide and phosphate fertilizer manufacture, but based on readily available information, the overall situation is summarized in Table 5.3.

With the exception of fuel and electricity, all these normally ignored emissions arise from intensive rather than organic farming. The good news is that arable land converted to mixed organic methods immediately starts taking carbon out of the atmosphere and putting it back in the soil. A recent UN report noted the 'substantial contribution of organic agriculture to climate change mitigation and adaptation' (Müller-Lindenlauf, 2009). This considered 11 trials comparing soil carbon levels on organic and non-organic farms and concluded that sequestration rates of up to 400kg carbon per hectare per year could be expected. A more detailed review of 39 studies for the Soil Association found that, on average, organic farms in Northern Europe contain 28 per cent more soil carbon than equivalent non-organic farms and calculated that they sequester on average 560kg (2053kg CO_2) per hectare per year for at least 20 years after conversion, with reducing amounts thereafter (Azeez, 2009).

Methane emissions from UK agriculture would increase if all farms were organic, because there would be more ruminants, but we would only be substituting animals grazed on land that is sequestering carbon, for

Table 5.3 *Hidden annual GHGs from UK agriculture in millions of tonnes (mt)*

Source	Total CO_2e (mt)
Fuel and electricity	2.4
N fertilizer production	8.5
Recent grassland conversion	9
Fens	1
Total	20.9

imported animals from land that has sometimes lost huge amounts of CO_2. According to Greenpeace International (2009), 80 per cent of rainforest clearance in the Amazon is to make way for cattle.

While this helps to justify an increase in ruminant numbers in the UK, it is still of interest to contrast the harmful effect of methane from ruminants with the soil carbon benefits of organic systems. Table 5.4 shows the annual CH_4 from non-organic beef and dairy cows. Emissions from organic beef cattle would be somewhat lower, since the carbon content of their manure is higher. Soils not receiving nitrogen fertilizer are also better able to break down atmospheric methane (Willison et al, 1995).

To express CH_4 in terms of CO_2 we need a conversion factor. The accepted IPCC figure of 21 does not cover the ongoing nature of ruminant emissions. To include this and recent research showing that the global warming potential of CH_4 has been underestimated, we actually need a multiply by about 90 (Young, 2009). The typical stocking rate for beef and sheep on organic mixed farms is 0.5 livestock units (LU) per hectare across the whole area. On all-grass organic farms it is about 1LU per hectare and on organic dairy farms between 1 and (exceptionally) 2LU per hectare.

By sequestering 560kg carbon per hectare, conversion to mixed organic farming using beef or sheep to provide fertility takes 2050kg of CO_2 out of the atmosphere each year (560 × 3.66). This is slightly lower than the CO_2e of the methane emission from the grazing animals at 2282kg. However, all arable farms in the UK use on average fertilizer containing 147kg of nitrogen per hectare each year. Since the production of each tonne adds at least 6.7 tonnes of CO_2e to the atmosphere, 147kg is responsible for 985kg CO_2e, giving a net advantage to the organic system of 753kg CO_2e per hectare each year. In relation to all-grass organic farms, where stocking levels are higher, the benefit is in preserving one of nature's major carbon stores. As we have already seen, ploughing a hectare of permanent grassland will release on average 2500kg of CO_2e every year for 100 years. Studies also indicate that established grassland sequesters carbon on an ongoing

Table 5.4 *CO_2 equivalent of the CH_4 emissions from beef and dairy cattle in the UK at various stocking rates*

	CH_4 exhaled/yr	CH_4 from manure/slurry/yr	Total CH_4/yr	CO_2e kg per cow	0.5 LU/ha	1 LU/ha	1.5 LU/ha
Beef cow	48kg	2.7kg	50.7kg	4563	2282	4563	
Dairy cow	105kg	25.8kg	130.8kg	11,772		11,772	17,658

Source: Methane data CEH (2008), p374

basis at about 200kg (733kg CO_2) per hectare per year (Müller-Lindenlauf, 2009). To this we can also add the benefit of not using nitrogen fertilizer of 985kg, a total of 4218kg, slightly short of the 4563kg from the CH_4 emissions.

The CH_4 emissions per cow from organic dairy cows will be about 30 per cent lower than those of intensive cows, principally due to better manure management from more welfare-friendly systems with more outdoor grazing and use of straw bedding, instead of slurry systems, but higher per litre of milk produced. Overall GHG emissions from all forms of dairying are relatively high, due to the high level of feed consumption. While this strengthens the case for us all to reduce our consumption of dairy products, we need to remember that dairy farms are highly productive, and, when integrated into a wider farming system can underpin the fertility needs of a large area of organic farm and horticultural land.

To challenge the claim that organic beef and lamb has a higher carbon footprint than chicken raised intensively indoors, we not only need to add in all the 'hidden' GHGs from ploughing and fertilizing the land on which grain is grown in the UK to feed the chickens, but also those from the enormous area of land outside the UK on which a significant proportion of concentrate feed is grown, such as the 2 million tonnes of soya used each year. This is a further omission of the Cranfield study. Data to undertake such a calculation completely is not available, so precise comparisons of different livestock systems cannot yet be made. A rough estimate based on what can be established suggests a possible small advantage still in favour of intensive chicken – if correct, a result of the phenomenally high growth and feed conversion rates achieved by intensive methods. As such, one question that arises is whether we can compare the nutritional value of intensive chicken with that of organic beef or lamb.

Grazing animals and human health

A major analysis by the World Cancer Research Fund (WCRF) of the sometimes contradictory scientific evidence concluded that high red meat consumption increases the risk of bowel cancer, while high white meat consumption does not (WCRF, 2007). Individual studies have found a link with prostate and pancreatic cancer and with low sperm count. Widely accepted concerns that the saturated fats in red meat increase the risk of heart disease have been included in the government's healthy eating guidelines. However, is it right to assume that these associations apply to all red meat, regardless of how it is produced? It has been pointed out that coronary heart disease was almost unknown in people under 60 during the early part of the 20th century, despite the fact that per capita consumption of red meat, lard, suet and butter was very much higher then than today (Enig, 2000). In the

UK average weekly consumption of beef fell from 244g per person in the early 1950s to 126g in the 1990s. In contrast chicken consumption during the same period increased from 19g to 237g (Defra, n.d.).

Part of the explanation for this apparent contradiction may lie in the difference between the way in which most farm animals were fed in the past and the way they are fed now. From the middle of the last century agricultural intensification caused virtually all chickens and pigs, plus a significant proportion of cattle, to be taken out of the fields and switched from the grass-based diet on which they had evolved, to one based on grain. In countries like the US and Australia this trend has been even greater than in the UK, with the rise of feedlots where thousands of cattle are kept on bare earth and fed a cereal and maize diet containing as little as 13 per cent fibre. In stark contrast the Soil Association's organic standards require that at least 60 per cent of cattle feed must be roughage such as grass or hay. It is now recognized that these factors affect the relative proportions of different fats in red meat and that this has implications for human health. A recent review of the published science has found that grass-fed beef has a lower total fat content, a higher level of stearic acid (a saturated fat not implicated in the rise of blood cholesterol levels) and lower levels of three fats that do increase cholesterol: palmitic acid, lauric acid and mystric acid. It also has higher levels of oleic acid, the fat present in olive oil. Even more importantly, levels of omega-3 polyunsaturated fats, known to protect against cancer and several other diseases, are up to 11 times higher in grass-fed than grain-fed cattle, while levels of omega-6 fatty acids are broadly the same in both types. As a result grass-fed beef has an average omega-6 to omega-3 ratio of 1.56:1. Grain-fed beef has an average ratio of 7.65:1. It is now known that too high a level of either the omega-6 or omega-3 group impedes the essential functions of the other, including the uptake of important fat-soluble vitamins. A diet in the range of 1:1 to 4:1 omega-6 to omega-3 is considered healthy, but typical Western diets today average 11:1 and can be as high as 30:1 (Daley et al, 2010). Research has shown that reducing this to 5:1 suppressed inflammation in patients with rheumatoid arthritis, while reducing it to 2.5:1 decreased colorectal cancer cells and the risk of breast cancer in women (Simopoulos, 2008). Meat and milk from predominantly grass-fed animals has other advantages too: higher levels of beneficial conjugated linoleic acid and of many other antioxidants and micronutrients.

Even the WCRF, which has led the campaign to reduce the consumption of red meat, acknowledges that the meat of wild animals has a very different fat profile to that of most farmed animals and may therefore not increase the risk of cancer (WCRF, 2007). Good organic systems and many of the herds and flocks in upland Britain, where little if any fertilizer is used are likely to produce meat with characteristics similar to that of wild animals.

In relation to chicken, scientists at the Institute of Brain Chemistry and Human Nutrition in London have highlighted the fact that more than half the energy in a modern broiler chicken (as well as some organic chickens) comes from fat, whereas 60 years ago, the vast majority of the energy came from protein. As with beef, levels of omega-6 fatty acids have not fallen, giving a highly unhealthy balance of almost ten times as much omega-6 as omega-3. The researchers associate this with the rise of dementia (Crawford et al, 2009).

The same study found that intensively produced chicken meat today contains only one-fifth the amount of the essential omega-3 fatty acid docosahexaenoic acid (DHA) found in wild birds. One of the authors of the study, Professor Michael Crawford, says: 'Essential fats for the brain are the priority... In biochemical terms the limiting factor for the brain is the omega-3 docosahexaenoic acid [DHA] ... to get the same amount of DHA from a modern broiler chicken you need to eat about 3 to 5 chickens at a cost of over £12 and with 5000 calories of thrombogenic and atherogenic fats included' (Crawford, 2009, personal communication).

Organic farming, biodiversity and animal welfare

The biodiversity benefits of organic farming have been recognized by many studies. As Alastair Rutherford, former head of agriculture at English Nature has said, 'Consumers can be confident that by demanding and buying produce from organic farms in England they will help reverse the declining fortunes of our farmland wildlife' (Anon, 2004). In Europe high animal welfare requirements form an integral part of organic standards. Sadly this is not yet the case in the US and some other countries.

Some conclusions

I have tried to show that reducing farming's carbon footprint is not just about methane. IPCC scientists have recognized that soil carbon sequestration represents 90 per cent of agriculture's climate-change mitigation potential and at the present time only mixed farming systems like organic have the ability to realize this while simultaneously producing adequate food crops.

When I played my part in ploughing up the Cotswolds 40 years ago, I put GHGs into the atmosphere that are still doing harm today. Organic farmers, of course, also plough grassland, but by keeping land in crops for only short periods before returning it to grass, and by carefully recycling nutrients through manures and compost, they manage to produce high yields in ways that are far less dependent on fossil fuels and other finite resources, while additionally retaining more carbon in the soil. If we change our diets and eat much less chicken and pork than we do today, slightly reduce our consumption of dairy products and eat beef (and lamb) only if the animals have been fed on grass and clover, then organic farming could feed present populations and possibly even the predicted 9 billion by 2050.

It is commonly assumed that the lower yields of arable organic farming in Europe mean it is not an option for feeding an increasing global population. Yield differences arise because subsidized European farmers can still afford to use very high fertilizer applications, despite recent dramatic price rises; African farmers generally cannot. In such countries the greater extremes of climate also mean that soil carbon levels (in the form of humus) have a much greater impact on crop yield than they currently do in Europe. Organic matter can hold 20 times its weight of water and improves the soil's sponge-like properties. It keeps plants growing during droughts and reduces soil erosion during heavy rain. It also helps crops to fight disease naturally. As such it is hardly surprising that a report by the Food and Agriculture Organization found that in Africa yields on organically farmed land were 140 per cent of those on non-organic land (UNCTAD, 2008).

Due to its reliance on grass and clover, more even than livestock, organic farming will always produce CH_4 emissions. But we need to see this in context – and more constructively than simply countering that less CH_4 comes from ruminants than landfill sites, with the kitchen waste from an average family producing almost as much CH_4 each year as two beef cows.

Having convinced myself that a broad analysis of agricultural GHGs demonstrates that on balance organic farming has the potential to help mitigate climate change, my mind turns to other aspects of the question posed by this chapter. On my way back from Somerset I stop to fill the Land Rover with fuel. The high price reminds me that when oil and natural gas are eventually in really short supply and nitrogen fertilizer becomes too expensive to manufacture, clover and other legumes will still be able to extract limitless quantities of nitrogen from the atmosphere to enrich the soil.

Phosphate fertilizers in contrast, the other essential input for a productive agriculture, are not an infinite resource. Research at Newcastle University has estimated that at current rates of usage, readily available sources will be all but exhausted within 60 years (Scott-Thomas, 2008).

Superphosphates, which I used before going organic, dissolve in water and bring a quick response in plant growth. British farmers use about 650,000 tonnes each year and they are the major cause of river eutrophication. They are made by adding acid to phosphate-rich rocks from the north coast of Africa. Rock phosphates, simply ground to a powder are used sparingly in organic farming. They become available in the soil more slowly because they are dissolved by weak acids from plant roots, not water.

But organic farming is not just a way of avoiding the harmful effects of dangerous chemicals; it is also a way of husbanding precious resources and avoiding disasters. The BSE crisis in the UK cost taxpayers over £4 billion. It is the reason for my long journeys to get animals slaughtered and it created immense hardship for the whole livestock industry. BSE has caused human tragedies too and so nearly a human health crisis of epic proportions. But future generations may look back and notice something even more fundamental.

By foolishly using MBM in the feed of naturally herbivorous animals, British feed compounders and farmers have deprived the pig and poultry sectors of what was a perfectly appropriate foodstuff for omnivores, and the agricultural industry of a natural fertilizer. MBM from cattle is no longer permitted in any feed or as a fertilizer because this could spread the prion proteins thought to be the infective agent of BSE. The loss of this resource is substantial. Millions of cows and millions of tonnes of MBM have gone into landfill or been incinerated. This has not only caused extra GHG emissions, it has also deprived agriculture of a major way of recycling the phosphate locked up in animal bones, increasing the need for phosphate fertilizers. In the not too distant future, when agricultural productivity is limited by lack of phosphates, the next generation may look back and see the BSE episode as symptomatic of a wider malaise affecting food production, which in so many respects still lacks common sense and epitomises the difference of approach between non-organic and organic methods. Organic farming is not perfect, but at least it seeks to address those big issues on which past civilizations have floundered and which still hang over us today.

As I pull back into the farmyard it is already getting dark and I still have a day's work to do, but at least I am now clear that I won't be ordering that fertilizer spreader after all.

References

Anon (2004) 'Organic farming increases biodiversity, says review', *Times Higher Education*, 13 October 2004

Azeez, G., (2009) 'Soil carbon and organic farming', Soil Association, Bristol, available at www.soilassociation.org/LinkClick.aspx?fileticket=SSnOCMoqrXs%3d&t abid=574, last accessed 7 June 2010

Carson, R. (1962) *Silent Spring*, Houghton Mifflin, Boston, MA

CEH (2008) 'Annexes of the UK Greenhouse Gas Inventory, 1990 to 2007', Centre for Ecology and Hydrology, Edinburgh, available at www.airquality.co.uk/reports/cat07/0905131425_ukghgi-90-07_Annexes_Issue2_UNFCCC_Final.pdf, last accessed 12 September 2009

CEH (2009) 'Inventory and projections of UK emissions by sources and removal by sinks due to land use change and forestry', Defra Contract GA01088, UK

Cormack, W. F. and Metcalfe, P. (2000) 'Energy use in organic farming systems', final report for project OF0182 for Defra, ADAS, Terrington

Cottingham, M. and Leech, A. (2009) 'Organic market report 2009', Soil Association, Bristol, p5

Crawford, M. A., Bazinet, R. P. and Sinclair, A. J. (2009) 'Fat intake and CNS functioning: Ageing and disease', *Annals of Nutrition and Metabolism*, vol 55, nos 1–33, pp202–228

Cross, J. (2009) 'EBLEX warning on beef industry decline', 31 May, available at http://store.eblex.org.uk/index.asp, last accessed 15 September 2009

Daley, C. A., Abbott, A., Doyle, P. S., Nader, G. A. and Larson, S. (2010) 'A review of fatty acid profiles and antioxidant content in grass-fed and grain-fed beef', *Nutrition Journal*, vol 9, no 10, pp1–12

Defra (2009a) 'General statistics on BSE cases in Great Britain', available at www.defra.gov.uk/vla/science/docs/sci_tse_stats_gen.pdf, last accessed 12 September 2009

Defra (2009b) 'UK emissions of all greenhouse gases, carbon dioxide, methane and nitrous oxide', Annex B National statistics – 2008 UK provisional figures, available at www.decc.gov.uk/en/content/cms/statistics/climate_change/climate_change.aspx, last accessed 28 September 2009

Defra (n.d.) 'National Food Survey 1950–2000', Defra, London

Doward, J. (2009) 'Milk "strikes" and shortages it Europe as UK dairy industry reels from crisis', *The Observer*, 20 September, p17

Enig, M. (2000) *Know Your Fats: The Complete Primer for Understanding the Nutrition of Fats, Oils and Cholesterol*, Bethesda Press, Bethesda, MD, pp84–112

Genesis Faraday (2008) 'Modelling the effect of genetic improvement on emissions from livestock systems using Life Cycle Analysis', Report for Defra, Appendix 2

Greenpeace International (2009) 'Slaughtering the Amazon', available at www.greenpeace.org/international/en/publications/reports/slaughtering-the-amazon/, last accessed 24 May 2010

Houghton, R. A. (1999) 'The annual net flux of carbon to the atmosphere from changes in land use 1850–1990', *Tellus*, vol 52, no 2, pp298–313

Indian Government (n.d.) 'Green revolution and its aftermath', Department of Soil and Water Conservation, Punjab, available at http://dswcpunjab.gov.in/contents/special_features.htm, last accessed 23 September 2009

Johnson, D. E. and Johnson, K. A. (1995) 'Methane emissions from cattle', *Journal of Animal Science*, vol 73, pp2483–2492

Jones, P. (2009) Interview on BBC Radio 4 Farming Today, 20 May, transcript available at www.self-willed-land.org.uk/articles/farming_today.pdf, last accessed 2 September 2009

Jones, P. and Crane, R. (2009) 'England and Wales under organic agriculture: How much food could we produce?' Centre for Agricultural Strategy, Reading

Kennedy, J. (2009) Letter from Rt Hon Jane Kennedy MP Minister of State for Farming and the Environment, to Tony Cunningham MP, 14 May

Leake, J. (2009) 'Burping lamb blows roast off menu', *The Sunday Times*, 24 May, p11, available at www.timesonline.co.uk/tol/news/environment/article6350237.ece, last accessed 24 May 2010

Melillo, Jerry (2009) 'Science in action', BBC World Service, 23 October

Mortimer, N. D., Cormack, P., Elsayed, M. A. and Horne, R. E. (2003) 'Evaluation of the comparative energy, global warming and socio-economic costs and benefits of biodiesel', Sheffield Hallam University, p28

Müller-Lindenlauf, M. (2009) 'Organic agriculture and carbon sequestration', FAO, Rome

Scott-Thomas, C. (2008) 'Wheat yields will be slashed as phosphate runs out, warns scientist', available at http://tinyurl.com/ykfrdar, last accessed 24 May 2010

Simopoulous, A. P. (2008) 'The importance of the omega-6/omega-3 fatty acid ratio', *Experimental Biology and Medicine*, vol 233, no 6, pp674–688

Steinfeld, H., Gerber, P., Wassenaar, T., Castel, V., Rosales, M. and de Haan, C. (2006) 'Livestock's long shadow: Environmental issues and options', FAO, Rome

UNCTAD (2008) 'Organic Agriculture and Food Security in Africa', UNEP-UNCTAD Capacity Building Task Force on Trade, Environment and Development (CBTF), United Nations Publication, New York and Geneva

Vellinga, Th.V, van den Polo-van Dasselaar, A. and Kuikman, P. J. (2004) 'The impacts of grassland ploughing on CO_2 and N_2O emissions in The Netherlands', *Nutrient Cycling in Agroecosystems*, vol 70, pp33–45

WCRF/AICR (2007) 'Food, nutrition, physical activity, and the prevention of cancer: A global perspective, World Cancer Research Fund/American Institute for Cancer Research, Washington, DC

Williams, A. G. (2007) Draft data from 'Developing and delivering environmental life cycle assessment of agricultural systems', Defra Project IS022

Williams, A. G., Audsley, E. and Sandars, D. L. (2006) 'Determining the environmental burdens and resource use in the production of agricultural and horticultural commodities', Main Report, Defra Research Project IS0205, Bedford, Cranfield University and Defra

Willison, T., Goulding, K., Powlson, D. and Webster, C. (1995) 'Farming, fertilizers and the greenhouse effect', *Outlook on Agriculture*, vol 24, no 4, pp241–247

Young, R. (2009) 'The role of livestock in sustainable food systems', Soil Association, Bristol, available at www.soilassociation.org/LinkClick.aspx?fileticket=qm0ueyxHQj I%3d&tabid=313, last accessed 24 May 2010

Part 2
Farming Practices and Animal Welfare

6

Food from the Dairy: Husbandry Regained?

John Webster

My neighbours, the Snell family, milk 120 cows in the south-west of England. This year the cows have been at pasture since February. Approximately 65 per cent of the food they eat, year round, has been grown on the farm. Our hamlet, Mudford Sock, and the adjoining villages Ashington and Lymington, are recorded in the Domesday Book. In 1086 they were described as pasture and meadow – and they still are today. By any environmental audit this must qualify as sustainable. Yet two of the six farms in our area have gone out of dairying in the last year. The national picture is just as dismal. According to the University of Manchester Centre for Agricultural, Food and Resource Economics, one in eight dairy farms in the UK has ceased production in the last two years.

Currently the Snells are being paid 22 pence/litre for their milk. In my local supermarket this retails for 67p/litre. A large bottle of fizzy water is priced at 96p/litre, 43 per cent more than the price of milk in the shop and over four times the price of milk at the farm gate. The crude logic of this is that we regard milk as a commodity whereas fizzy water in a green bottle is treated as an added value product and we pay through the nose for the bubbles. So long as the suppliers of the supermarkets can obtain milk at 17.4p/litre that is the price they will pay. The invisible hand of the free market writes that this is good for us all because we are all consumers. If farms die, this is of no more consequence to the common good than the loss of the miners or workers in the motor industry.

There are two serious flaws in this *laissez faire* argument that impact not just on the farmers but on us all. It affords no value to the quality of the living countryside and it affords no value to the welfare of the dairy cow. We, especially those of us who drink bottled water, may pay lip service to the environment and farm animal welfare; we express a desire for higher standards, but this desire is seldom matched by demand. The real measure of our demand for adding value to the quality of the countryside and the quality of animal welfare is the price we are prepared to pay for these things. If we valued them as highly as fizzy water we would be prepared to pay twice as much.

The essence of the problem is that milk, when it leaves the farm, is viewed as no more than a commodity. Others in the supply chain may convert this milk into a highly attractive range of added value products, such as fruit yoghurts and regional speciality cheeses, and make good money in the process. A few dairy farmers can add value on the farm, e.g. by producing and marketing their own cheese or ice cream. The cow, however, can do no more than produce the basic commodity, milk; so is trapped at the bottom of the supply chain. In consequence she has become the most extreme manifestation of exploited labour. The high-yielding dairy cow is, by some distance, the most hard worked of any animal (including human animals) and it is little surprise that so many succumb to a number of production diseases and have to be culled after a working life that can be nasty, brutish and far too short to make sense even by the crudest of economic measures.

How did we get where we are today?

How came this sorry state of affairs? Within cultures that developed in close association with cattle as chattels for their personal use (the two words have the same root), the most valuable of non-human animals has been the lactating cow. She has been a daily provider of food in the form of milk to be consumed at once or conserved as butter, cheese, ghee or yoghurt. Not only is milk a highly nutritious food for the family, it is also available for sale on a daily basis and thus a regular source of income. Farmers and their families invested time and labour to grow, cut, carry and conserve food for the house cow, and the cow more than repaid this investment through the production of milk for home consumption and for sale. Throughout history and worldwide, dairy production has offered those who work the land a currency that could lift them from subsistence agriculture towards some freedom of choice in the marketplace. At best therefore, a family could thrive on the wealth provided by their milch cow or cows. At worst, if they faced ruin through poverty, age or illness, they went down together.

Dairy farming in the developed world has moved a very long way from the position where the individual cow was a valued member of the family. In the developed world of industrialized agriculture, herds of more than 100 animals are the norm and massive dairy factories in which cows are permanently confined in herds of over 1000 animals are becoming increasingly common. In the last 30 years annual lactation yields of the most productive herds of dairy cattle have increased from less than 5000 to over 10,000 litres. These total yields correspond to peak daily yields of approximately 25 and 50l/day. This has been achieved by parallel processes of genetic selection for increased yield, and changes in diet and feeding practice designed to increase nutrient supply.

It would be lazy (and dishonest) to assume that intensification of dairy production leads inevitably to abuse of the principles of good husbandry, whether measured in terms of animal welfare or environmental quality. Indeed some of the worst abuses of welfare (both animal and human) can arise on small family farms from problems that are as old as agriculture, namely failure to provide through poverty, age and illness; in short, failure to cope. Cows in large commercial herds adequately serviced by people and machines are less likely to suffer from catastrophic failures of provision. There is, however, quite enough evidence to show that cows in large commercial units can suffer from systematic failures to provide the quality of husbandry necessary to meet each and every one of the Five Freedoms:

1 Freedom from Hunger and Thirst: by ready access to fresh water and a diet to maintain full health and vigour
2 Freedom from Discomfort: by providing an appropriate environment including shelter and a comfortable resting area
3 Freedom from Pain, Injury or Disease: by prevention or rapid diagnosis and treatment
4 Freedom to Express Normal Behaviour: by providing sufficient space, proper facilities and company of the animal's own kind
5 Freedom from Fear and Distress: by ensuring conditions and treatment which avoid mental suffering.

Source: Farm Animal Welfare Council (2009)

Some of the most important failures of provision are outlined in Table 6.1 and explained within the following bullet points:

*   She may both suffer and fail to sustain fitness through hunger, malnutrition or metabolic disease if she is unable to consume or digest sufficient nutrients to support her genetic and physiological potential to produce milk.
*   She may suffer chronic discomfort if housing design, especially the design of her lying area, is inappropriate to her size and shape. Problems of poor cubicle design and inadequate bedding may become worse if she loses condition through malnutrition.
*   She may suffer pain through lameness or mastitis.
*   She may be at greater risk of infectious disease either through increased exposure to infection or increased susceptibility in consequence of metabolic stress.
*   She may be bullied or denied proper rest by other cows. She will, almost inevitably suffer the loss of her calf shortly after birth.

Table 6.1 *Abuses of the 'Five Freedoms' that can arise through systematic failures in the provision of good husbandry*

Hunger	Nutrition fails to meet metabolic demands for lactation
Chronic discomfort	Poorly designed cubicles, inadequate bedding
Pain and injury	Claw disorders (sole ulcer, white line disease) Damaged knees and hocks Mastitis
Infectious disease	Mastitis, digital dermatitis
Fear and stress	Rough handling, bullying, separation of calves at birth
Suppressed behaviour	Zero grazing, inadequate resting time
Exhaustion	Emaciation, infertility, premature enforced culling

These potential sources of poor health and welfare can be interdependent and additive. For example, the high genetic merit dairy cow, housed in cubicles and fed a diet based on wet grass silage and concentrate (cereal and soya type feeds), may suffer both from hunger and chronic discomfort, partly because the quality of feed has been inadequate to meet her nutrient requirements for lactation and she has lost condition, partly because the wet silage has contributed to poor hygiene and predisposed her to foot lameness and partly because genetic selection has created a cow too big for the cubicles.

The Five Freedoms have proved to be an exceptionally valuable tool for assessing the welfare state of an animal, or a herd of animals, at a moment in time. However, for many farm animals, and for dairy cows in particular, some of the most severe welfare problems arise from the long-term consequences of trying, and ultimately failing, to cope with the exacting physiological and behavioural demands of everyday life. Exhaustion, the final welfare concern listed in Table 6.1, is not identified within the Five Freedoms but is probably the biggest welfare problem of all. It describes the cow broken down in body, and probably in spirit, through a succession of stresses arising from improper nutrition, housing, hygiene and management, exacerbated in many cases by inappropriate breeding due to selection for production traits at the expense of fitness. She is likely to be emaciated and infertile. She may be chronically lame. Such cows will be culled because they are no longer making a productive contribution to the enterprise. The real shame is that far too many cows are culled far too young (after three lactations or less). This is not only an abuse of welfare but also a terrible waste since a dairy cow needs to complete at least four lactations to recoup the cost of rearing her as a heifer to the time that she delivers her first calf and enters the milking herd.

Although some welfare problems for dairy cattle may arise from bad luck, most may be attributed to systematic failures of provision. Thus the three major problems of health and welfare, namely infertility, mastitis and lameness, are all described as 'production diseases', namely diseases and disorders arising as a direct consequence of the production system imposed on the animals. In order to address these problems it is necessary to identify the hazards implicit within the system, to assess the magnitude of the risks arising from these hazards, to develop a strategy to address these hazards (a herd health and welfare plan) and to ensure that this plan is carried out to good effect.

Cows at the limit

So long as food from farm animals is regarded by us, the general public, simply as a commodity, the only mechanisms whereby farmers can sustain their own living and their own farms in the face of international competition are through increasing productivity and reducing costs. The dairy industry has responded to this challenge with a policy designed to maximize milk yields from individual cows. This policy has been justified on the following grounds:

- increased feed conversion efficiency (relatively more feed is converted into milk relative to that required for the cow's bodily maintenance);
- increased income from milk sales relative to the fixed costs of labour, housing and machinery (it costs the same to milk a cow yielding 5 or 50l/day);
- a potential reduction in environmental costs (less production of carbon dioxide and methane relative to milk yield).

As indicated earlier, these huge increases in milk yield per cow have been achieved through a combination of genetic selection for increased yield (especially in the Holstein breed), changes in diet necessary to meet the ever increasing metabolic demands of lactation and changes in housing and management designed to speed up the daily chores of feeding, milking and mucking out, reduce labour costs and ensure that cows spend as much time as possible in productive work (taking in feed and yielding milk).

No farm animal is worked harder than the dairy cow. The proof of this contention is provided by Table 6.2, which compares the daily food energy requirements and energy expenditure (in other words, work, expressed as heat production) of a range of humans and farm animals. To simplify the comparison, values for energy intake and heat production are expressed in relation to that of an adult sedentary human (e.g. a typical office worker or

Table 6.2 How hard do animals work?

Species	Activity	Energy exchange		
		ME intake	Work/heat	'Food'
Human	Sedentary	1.00	1.00	
	Working miner	1.25	1.25	
	Lactating woman	1.53	1.28	0.25
	Endurance cyclist	2.60	2.90	−0.30
Pig	Grower	2.10	1.30	0.80
	Lactating sow	3.20	1.73	1.47
Birds	Broiler chicken	2.10	1.18	0.92
	Laying hen	1.73	1.30	0.43
	Passerine feeding chicks	3.03	3.03	
Cow	Suckler with one calf	2.22	1.32	0.91
	Dairy cow, 50l/day	5.68	2.14	3.53

Note: This table compares the daily food energy requirements (expressed as Metabolizable energy, ME), energy expenditure as work (heat production), 'food' energy outputs, expressed as weight gain in meat animals, eggs or milk. The sedentary adult human (e.g. office worker) is taken as the basis for comparison (ME intake and work output as heat = 1.0). Energy exchanges of all other classes are expressed as multiples of this standard sedentary human.

university student). (For a more complete explanation of this comparison based on the concept of metabolic body size (weight, kg$^{0.75}$) see Webster, 2005, p133) Relative to an adult sedentary male, a lactating mother will eat 38 per cent more food energy and work 13 per cent harder; 25 per cent of the food energy she consumes will be carried into her milk. Miners digging coal by hand worked about 25 per cent harder than sedentary individuals, but slightly less than the mothers of their children. Only those engaged in extreme sports such as the Tour de France work to their absolute limits, and then not on a full-time basis. Growth in farm animals is not a particularly energy demanding process. Even in the rapidly growing broiler, heat production is less than 20 per cent above maintenance. The laying hen, producing one egg per day, is faced with considerable metabolic demands, e.g. in relation to calcium metabolism. However, the energy cost of egg-laying is not particularly severe (heat production 30 per cent above maintenance). All these costs pale into insignificance when set against the cost to a dairy cow of sustaining a milk yield of 50l/day. She has to consume an amount of feed nearly six times that required for bodily maintenance and her work load (heat production) exceeds twice that of maintenance. The demands of lactation for a sow feeding ten piglets (or a Labrador bitch feeding eight puppies) are also high. However, they do not approach that of the high yielding dairy cow. Moreover they are unlikely to persist longer than about eight weeks, by which time the sow or bitch will almost certainly have lost a lot of condition. The dairy cow has the added problem of sustaining the metabolic demands of lactation for most of her working life. The only animal species that I have discovered to work harder than dairy cows are birds while feeding their young in nests. This too is a process that doesn't last too long. It is little wonder that dairy cows break down with signs of exhaustion.

Synthesis of so much milk not only presents the dairy cow with an enormous metabolic load, the need to consume enough feed to meet this metabolic demand (five to six times that required for maintenance) drives the digestive system to its limits and can seriously compromise their need to rest and sleep.

A typical Holstein cow grazing excellent pasture can only consume enough grass to sustain a yield of about 25l/day. If a cow is to sustain a yield of 50l/day or more she must have near continuous access to a highly nutritious diet (a Total Mixed Ration, TMR) containing a high proportion of mechanically harvested cereals and protein supplements such as soya. At extremely high yields, this makes it unprofitable to turn cows out to pasture where they simply cannot take in nutrients fast enough. This then leads to the practice of zero grazing, whereby cows are confined through most or all of lactation and may be allowed out to pasture (if at all) during a period of about two months at the end of lactation and before the birth of their next calf.

However well-formulated the diet for high yielding dairy cows, metabolizable energy (ME) demand in early lactation exceeds that which can be achieved by the cow within the constraints of appetite and the cow loses condition. She 'milks off her back'. The physiological demand for nutrients to support lactation creates a condition of 'metabolic hunger' and the intensity of this hunger increases as body condition falls. The loss in body condition and increase in hunger is more extreme when cows are injected with bovine growth hormone (BST) to stimulate increased milk yield (Chalupa and Galligan, 1989).

It is salutary to consider these stresses on the physiology of digestion and metabolism in terms of how it feels to the high yielding cow. She is motivated to eat by metabolic hunger (a function of both milk yield and body condition). She is motivated to *stop* eating by sensations (conscious or unconscious) associated with gut overload and the conflicting desire to do something other than eat, such as rest. At pasture, the first constraint on food intake is the rate at which the cow can physically consume the grass. Here the motivation for the cow to stop eating is most likely to arise from the desire to rest than from the sensation of gut fill. Very high yielding cows confined within a zero grazing system are presented with the conflicting problems of metabolic hunger, gut overload and the need to rest in a more extreme form and face a more difficult compromise as they seek to minimize the discomfort involved in reconciling these three stresses.

This summary of the conflicts between metabolic need, digestive capacity and the need for rest provokes a number of difficult practical questions concerning dairy cow welfare.

- Is it more stressful to a dairy cow to produce 55l milk/day than 25l/day?
- Is it stressful to restrict a cow with the genetic potential to produce 55l/day to a diet that can only sustain 25l/day?'
- Is it stressful to restrict a cow to a zero grazing system for most or all of her adult life?

There are no easy answers to these questions. The use of high genetic merit Holsteins (60l/day potential) has presented welfare problems manifested most obviously by infertility and early forced culling (Pryce et al, 1997). This is particularly evident on pastoral systems (e.g. in New Zealand) where ME intake cannot keep pace with metabolic demand (Harris and Winckleman, 2000). The best dairy farmers operating this system are now working towards a better balance between supply and demand both by improving the quality of the pasture and by selection within the Holstein breed, or through cross-breeding to produce a more robust cow with a lower potential yield but an improved lifetime performance.

The strategic decision to maximize milk yield per cow almost inevitably imposes the need to provide lactating cows with continuous access to a highly nutritious, premixed TMR. This is incompatible with a system that allows cows to have daily access to pasture. Benefits, other than fresh grass, that cows may derive from access to pasture include improved comfort, opportunities for exercise and social behaviour. It is possible to address these needs within intensive, zero grazing systems, but they are seldom achieved to the cows' satisfaction.

'Unfit for sustained work'

Rather less than half the 'high genetic merit' dairy cows in the developed world currently manage to sustain a working life of more than three years (three lactations). The reasons for this are complex and, once again, there are no simple solutions. The three main causes of enforced culling (removal of unproductive cows from the herd) are infertility, mastitis and lameness (Esslemont and Kossaibati, 1996). As stated earlier, these may all be considered as 'production diseases' for which the major risks arise from hazards intrinsic to the system and management. A major hazard for infertility is loss of body condition associated with inadequate nutrition in early lactation (Collard et al, 2000). Mastitis is caused by infections with bacteria arising from the environment or sequestered within the udder of carrier cows and transmitted at the time of milking. The costs to the farmer of mastitis are obvious and immediate: discarded milk, costs of treatment and loss of production in the longer term. Consequently most farmers operate a strict mastitis control policy through attention to good hygiene and the elimination of infected carrier cows with high somatic cell counts (SCC). In many countries this strategy is rewarded through an economic policy of carrot and stick: increased milk price for low SCC, financial penalties when SCC in the bulk milk tank is high.

Lameness is the most severe welfare problem for the dairy cow by virtue of the pain involved and its prevalence. On random inspection of the milking cows on a modern commercial dairy farm over 25 per cent, on average, are likely to show visible signs of lameness (Cook, 2003; Whay et al, 2003). This too is a complex problem involving a multiplicity of hazards associated with housing, hygiene, nutrition, breeding and management. Moreover there is less incentive to control lameness than mastitis, because the economic costs are less apparent and usually there is no financial incentive to take remedial action (but see below). Many farmers have assumed fatalistically that some degree of lameness is an inevitable consequence of a system that involves concrete floors and lots of slurry. However, much recent research has consistently revealed that the most important risk factors for lameness are associated with failures of foot care, improper or non-existent foot trimming and failure to treat new cases at the first sign of abnormal locomotion (Smith et al, 2007; Bell et al, 2009). One can sympathize with farmers and stockpeople who are so hard worked that they find little time to attend to cows' feet. However, they should be aware that if they do, things will get better. There is no excuse to adopt the defeatist approach and claim that there is no cure for lameness short of rebuilding the entire farm.

While these three widespread production diseases can be attributed mainly to failures of nutrition, housing and management, there is no denying that things have been made worse for both farmers and their cows through a breeding policy that selected cows overwhelmingly on the basis of increased productivity (milk yield in first lactation) without sufficient attention to fitness traits such as fertility, resistance to mastitis, sound locomotion etc. that would reduce their susceptibility to the major production diseases and enable them to sustain a longer working life, while feeling better in the process.

Table 6.3 (from Pryce et al, 1998) describes the phenotypic and genotypic correlations between selection for increasing milk yield, calving interval (a measure of infertility), mastitis and lameness in UK Holstein/Friesian

Table 6.3 *Phenotypic and genotypic correlations between milk yield and three indices of fitness in the dairy cow*

	Phenotype	Genotype
Calving interval	+0.20	+0.39
Mastitis	−0.01	+0.26
Lameness	+0.04	+0.17

Source: Pryce et al (1998)

cows. The phenotypic correlation describes the association as it appears on farm, the genotypic correlation that which is attributable to breeding rather than management and this is positive and highly significant in all cases. The negligible phenotypic effects on mastitis and lameness reveals that farmers are just managing to keep things under control through improvements to husbandry designed to compensate for the genetic deterioration in fitness. In the case of infertility, the battle is being lost.

Fortunately for all, dairy farmers and breeders are becoming increasingly aware of the need to modify selection indices to produce a more robust cow. Whereas ten years ago over 75 per cent of selection pressure would typically be directed towards production *per se* (yields of milk protein and fat), modern selection indices give increasing emphasis to traits linked to robustness and leading to improved lifetime performance. Indeed within Holstein UK, fitness traits now contribute just over 50 per cent to the selection index. This is undoubtedly a step in the right direction though it will take at least five years for its effects to become apparent.

Complementarity

There can be no doubt that the industrialization and intensification of dairy production has pushed dairy cows to their limits without bringing any obvious benefit to dairy farmers. The sheer size of the industry has also placed great strains on the environment, mainly through the destruction of forests and permanent pastures to create vast tracts of maize and soya bean grown for livestock feed. Moreover, it is certain that the size of the international dairy industry will continue to grow as increasingly affluent Asian consumers come to discover the delights of milk and milk products. Ruminants also produce large amounts of methane as a natural consequence of fermentation of carbohydrates in the rumen and the global warming effect of methane is 21 times that of carbon dioxide. These issues are presented in a scholarly, balanced, yet truly alarming way in Chapter 3 by Tara Garnett.

However, before one dismisses dairy production, along with meat production, as another glaring example of man's inhumanity to other animals and to the sustainable environment through flagrant misuse of resources, especially the consumption of food that would better go to feed the human population, it is useful to reflect on the extent to which animals in modern farming (as in traditional farming) can subsist on food that is complementary to, rather than in competition with, our own needs (grass and forages, by-products of foods designed primarily for human consumption, biofuels etc.). Table 6.4 compares efficiencies of production (output/input) of four of our major foods of animal origin, eggs, pork, milk and beef, when output is measured in terms of food energy and protein for man, and input is measured in terms of overall food energy, energy input from fossil fuels and, finally, feed energy input in terms of output relative to food directly available to man. Production of milk and eggs is inherently more efficient than meat production as a source of protein for humans because it doesn't involve killing the animals. Beef production from cattle grazed extensively on range is very inefficient in terms of overall feed energy use, but the most efficient in terms of fossil fuel use (although even this does not break even). Milk production is by far the most efficient when measured in terms of food energy and protein supply to humans, relative to consumption of food directly available to man. This is because a typical well-balanced ration for dairy cows can be formulated (and, in the UK, typically is formulated) on the basis that over 75 per cent comes from feed that is complementary to, rather than in competition

Table 6.4 *Energy use in animal farming*

	Hen	*Sow*	*Dairy cow*	*Range beef*
Primary yield per annum	300 eggs	1300kg pork	6000 l milk	200kg beef
Allocation of feed energy (kJ/MJ)				
to productive animals	790	830	720	580
to 'support' animals	210	170	280	420
Efficiency of yield				
kJ/MJ total feed energy	140	182	170	37
g protein/MJ feed energy	2.31	1.09	1.70	0.16
kJ/MJ fuel energy	210	265	136	550
MJ/MJ food for man	0.25	0.35	1.76	0.35
g protein/g protein food for man	0.37	0.19	1.60	0.14

Note: The two bottom rows describe the yield of energy and protein as meat, milk or eggs, relative to the animals' consumption of food that could have been directly fed to humans (for full explanation, see text).

with, the needs of man. In the example provided by Table 6.4, milk energy and protein yields are 1.76 and 1.60 energy inputs from 'food for humans'. By these criteria, the dairy cow becomes the most sustainable of farm animals, more so than the hen. I acknowledge at once that this (like all the others) is an incomplete argument but it is more comprehensive than many. One cannot condemn cows purely on the evidence of methane.

Husbandry regained

Whatever some may wish, dairy production is not going to go away. It follows therefore that our responsibility is to do it better. Here again, 'doing it better' does not necessarily imply, as some may claim, going back to traditional farming nor, as others will assert, continuing to develop down the same intensive route. We need to do both and we need to do both better.

I offer suggestions for action in four areas:

- improved husbandry in intensive dairy systems;
- development of pastoral 'free range' dairy systems as an integral part of stewardship of the living environment;
- improved, independent procedures for audit and implementation of improvements to health and welfare;
- promotion of value-added dairy products where quality of life (for cows and the living environment) is recognized as an essential measure of value.

Improved husbandry in intensive dairy systems

I offer the following six suggestions as practical steps towards improved husbandry within intensive dairy systems, defined both in terms of cow welfare and environmental quality.

1 Improvements to the provision and formulation of diets through better use of digestible fibre and by-products from foods grown principally for human consumption (e.g. sugar beet pulp) or biofuels (e.g. high protein oilseed residues). The aim should be to make at least 75 per cent of a Total Mixed 2 Ration for zero grazed dairy cows from food sources that are complementary to, rather than competing with the food needs of humans.
2 Improvements to building design designed to give cows more comfort when lying down (better beds), freer movement, especially when changing position from standing to lying (better cubicles), enriched, sheltered loafing areas to permit more exercise and social behaviour and softer, less abrasive walking surfaces, to reduce the risk of foot injuries.

3 Better attention to hygiene and management, especially foot care and early treatment of lame cows.
4 Reduction of pollution and energy costs, e.g. through the use of anaerobic digesters to ferment manure and other wastes and capture methane as an energy source.
5 Improved lifetime performance and welfare through increased emphasis on selection for robustness relative to milk yield.
6 Reduced production of male calves of extreme dairy type that are unfit for beef production and are condemned to immediate slaughter or (which is worse) intensive veal production. This can be encouraged partly through the used of sexed semen and also by increasing the productive life of the cows. A cow that produces six calves may deliver four of these calves to a beef-type bull. If she is likely to produce three calves or less, the farmer may feel compelled to use a dairy-type bull for every insemination.

Development of pastoral 'free range' dairy systems

There is a real alternative to the further development of intensive, probably zero grazed systems based on the modern high yielding Holstein cow (or her more robust descendants). This will be a development of (rather than a reversion to) sustainable pasture-based systems such as those that evolved where I live in the south-west of England and which I described in the opening to this chapter. We have seen a massive development of milk from grass in New Zealand, mainly as a source of butter and cheese for export. Despite advertising claims that this system is 'free range' it is less idyllic than it may appear. Herds in excess of 500 animals are being kept on monocultures of grasses and walked very long distances to and from the milking parlour. These large herds are creating new problems for the cows, especially increased prevalence of lameness, and for the environment, through destruction of mixed, sustainable ecosystems.

An environmental audit of dairy farming systems designed solely to produce income from the sale of milk and milk products will show that pasture-based systems operating at maximum output pose a greater potential threat to the environment (per unit of milk produced) than intensive, zero grazing systems through such things as increased nitrogen pollution, increased production of methane (a potent greenhouse gas) and increased fossil fuel consumption for the production of artificial fertilizers. If dairy cows are to be let out into fields, rather than confined in milk factories, then the best way to realize the principles of good husbandry is to accommodate dairy production within all the other elements that contribute to sustaining the quality of the living countryside. There is a real place for extensive, lower output, dairy farming within a broader definition of sustainable agriculture, where income from food production makes a significant but not dominant contribution to the cost of sustaining the living environment. Dairy cows at pasture have made an invaluable contribution to the sensitive and sustainable stewardship of the beautiful countryside in Britain and elsewhere. Where I live, the cows are moved through small fields with high hedges which are a Mecca for wildlife and the lanes between the fields are an explosion of wild flowers. When the cows come in for milking, they bring in the muck but the muck attracts the insects and the insects attract the swallows and martins, who nest in the eaves and the swifts who soar, and sleep, in the sky. If this is what the urban majority wish to preserve for our own peace of mind, we must be prepared to acknowledge that it is worth more than the price we pay for milk, either directly or via taxation through redirection of agricultural subsidies. The Common Agriculture Policy (CAP) requires farmers to comply with certain animal welfare and environment legislation as a condition of receiving their subsidies. The CAP also allows the European Union countries to make additional payments to farmers for actions to improve environmental quality and animal welfare, but there has been disappointing take-up by many countries of the option to make payments to improve animal welfare.

Audit and implementation of improvements to health and welfare

Nearly all farming systems within the UK, and most within the developed world, are subject to some form of audit procedure that purports to set standards of husbandry, hygiene and welfare that can be incorporated into a quality assurance scheme operated by retailers, producer cooperatives or non-government agencies such as the organic standards of the Soil Association and the RSPCA 'Freedom Food' scheme. The aim of these schemes is to assure consumers that the food they buy meets the standards they should expect in terms of safety, quality, provenance and other elements of quality such as animal welfare. At the time of writing most of the welfare assessment protocols operated by quality control agencies such as the UK Assured Dairy Farms (ADF) scheme (NDFAS, 2010) are based almost entirely on audit of the provision of resources to the animals and records of management procedures, such as the provision of health care. In a strictly practical sense, this approach has the merit that the information can be collected in objective fashion, free of observer bias. However, it is now generally recognized that these observations and records of the provisions necessary to establish good husbandry should be augmented, and in many cases replaced, by animal-based measures that provide a more direct assessment of animal welfare. This principle has been embraced by the European Commission, which has funded a major integrated programme 'Integration of animal welfare in the food quality chain: from public concern to improved welfare and transparent quality' (acronym WELFARE QUALITY, available at www.welfarequality.net). The main aims of this programme are to develop and implement robust on-farm welfare monitoring and information protocols for selected farm animal species. The approach has been described by Botreau et al (2007) and is based on the four principles of good welfare: good feeding, good housing, good health and appropriate behaviour. In essence, these are a more prosaic restatement of the Five Freedoms. Each principle is defined by a set of welfare criteria: e.g. good housing is defined by comfort at rest, thermal comfort and ease of movement. Each of these criteria may be established from a more comprehensive series of assessment procedures (observations and records) that may be conducted on farm. This strategy closely follows the development of the Bristol Welfare Assurance Programme. Its application to the audit of welfare in dairy herds has been described by Whay et al (2003).

The Bristol protocol for audit of the welfare of dairy cows was designed to incorporate the principles of good feeding, good housing, good health and appropriate behaviour and assess them in terms of observations and records that would be robust and reflect long-term consequences, rather than snapshots of welfare state (e.g. hock injuries rather than state of cubicle bedding on the day of inspection). This is not the place to review this study in detail but one observation was of special note. The overall lameness prevalence was 23 per cent, with a range extending from 0–14 per cent in the top 20 per cent of herds to 30–50 per cent in the bottom 20 per cent. However, farmer estimates of lameness prevalence in their own herds were, on average only about 20 per cent of the true figure. Weary farmers limping alongside their weary cows come to accept as normal that which is neither natural nor healthy.

We are satisfied that it is possible to carry out an honest, robust, independent animal-based audit of dairy cow welfare within a single day. The problem arises in carrying forward this audit into an effective dynamic programme for the delivery of sustained improvements to cow welfare, sustained rewards to farmers who promote improved welfare and sustained increase in consumer demand for added value dairy products, where quality of life for the cows and for the living countryside are recognized as indices of value.

To illustrate my approach to the delivery of improved husbandry and welfare through simultaneous programmes of improvement on farm and promotion through the food supply chain I have proposed the concept of the 'Virtuous Bicycle' (Webster, 2009), which is illustrated in Figure 6.1. Each wheel describes a virtuous cycle of assessment, action and review, but neither can operate without the other. The accreditation body sets husbandry and welfare standards acceptable to both producers and consumers/retailers. The sequence of events for the producer (the 'on-farm wheel') is as follows. The farmer first carries out a self-assessment of the enterprise to check on compliance with standards and identify any problems. An independent monitor then assesses the unit using an animal-based protocol looking mainly at welfare outcomes. Farmer, monitor and veterinary surgeon then address any immediate problems and devise a living strategy for health and welfare. The effectiveness of this strategy is reviewed after an appropriate time (e.g. one year or less if there are problems that need to be resolved quickly). The effectiveness of the strategy then feeds back to the farmer for further self-assessment and to the accreditation body who can benchmark the farm against approved standards and provide real assurance to the public as to what is being done. The second virtuous cycle, operating at retailer level, involves a process of information transfer between the accreditation authority and the public that sets out clearly the quality standards and provides honest evidence to indicate how well the scheme is working.

This concept of the 'Virtuous Bicycle' as a delivery vehicle for improved farm animal welfare emerges from an awareness that current welfare-based quality assurance schemes have a long way to go to achieve their joint aims of significant improvement in animal welfare at farm level and significant increase in consumer demand for proven high-welfare food. For these things to occur in practice, a quality assurance scheme must be seen

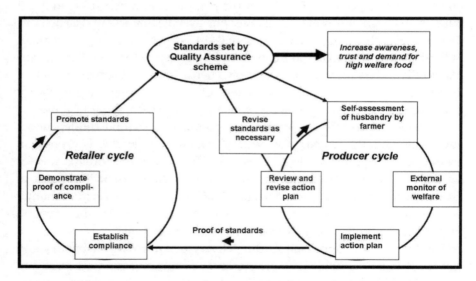

The producer cycle illustrates a dynamic process of self-assessment, external monitoring, action and review on farm. The retailer cycle describes the process of quality control and quality assurance at the retail level.

Source: Webster (2009)

Figure 6.1 The Virtuous Bicycle: A vehicle designed to deliver improved animal welfare on farm

to bring rewards to all those involved in the process: consumers, retailers, farmers and, of course, the animals. Farmers are unlikely to buy into the scheme without the assurance that it will bring them rewards in the form of increased income and security of contract. Consumers are unlikely to pay more and retain faith in the scheme unless they can trust the evidence upon which the assurance is based. In short, the Virtuous Bicycle will only deliver when both wheels turn together.

Dairy produce: What is its real value?

Milk from cows, like eggs from unmated hens, is food, pure and simple. It has been designed to feed the calf, after birth, or the chick before hatching, with a rich and balanced supply of nutrients. When humans first elected to harvest this food for our own consumption, we had no knowledge of subtle species differences in nutrient requirements, nor potential health risks associated with consuming too much rich food for too long. We ate it and thrived, and the benefits were most marked when small amounts of these foods of animal origin were added to an otherwise impoverished diet. The classic study in which John Boyd Orr demonstrated the benefits of adding milk to the diet of Scottish children in the 1920s led to the free school milk policy in Britain (not always appreciated at the time by children such as me). Dried milk products are a vital element of famine relief. While a minority will always reject dairy products on ethical grounds and another minority will avoid dairy products on health grounds (inescapable allergies or self-imposed obesity and the diseases of obesity), it is an inescapable fact that most people drink milk and eat butter, cheese, yoghurt and ice cream because they like it; and, if not consumed to excess, it is good for them. We are less likely to enrich the life of cows, farmers and especially the green and pleasant pastoral countryside by arguing that milk is unhealthy and cruel than by encouraging greater awareness of the real value of dairy produce. This should first recognize that highly nutritious, highly attractive food can make an immense contribution to our own health, welfare and enjoyment of life when consumed in moderation, but is dangerous when carried to excess. At the same time we should recognize that dairy produce is more than just a commodity. It is produced by dairy cows, sentient and highly engaging creatures, who share (to some degree) our ability to recognize quality in life, but are at the moment not getting a fair deal and deserve better.

References

Bell, N. J., Bell, M. J., Knowles, T. G., Whay, H. R., Main, D. C. J. and Webster, A. J. F. (2009) 'The development, implementation and testing of a lameness-control programme based on HACCP principles and designed for heifers on dairy farms', *The Veterinary Journal*, vol 180, pp178–188

Botreau, R., Veissier, I., Butterworth, A., Bracke, M. B. M. and Keeling, L. (2007) 'Definition of criteria for overall assessment of animal welfare', *Animal Welfare*, vol 16, pp225–228

Chalupa, W. and Galligan, D. T. (1989) 'Nutritional implications of somatotropin for lactating dairy cows', *Journal of Dairy Science*, vol 72, pp2510–2524

Collard, B. L., Boettcher, P. J., Dekkers, J. C., Petitclerc, D. and Schaeffer, L. R. (2000) 'Relationships between energy balance and health traits of dairy cattle in early lactation', *Journal of Dairy Science*, vol 83, pp2683–2690

Cook, N. B. (2003) 'Prevalence of lameness among dairy cattle in Wisconsin as a function of housing type and stall surface', *Journal of the American Veterinary Medical Association*, vol 223, pp1324–1328

Cross Compliance (2010) available at www.crosscompliance.org.uk/cms/, last accessed 24 May 2010

Dore, C. J. (and 17 others, 2007) 'UK emissions of air pollutants', 1970–2006 National Agricultural Engineeering Institute, available at www.naei.org.uk, last accessed 24 May 2010

Esslemont, R. J. and Kossaibati, M. A. (1996) 'Incidence of production diseases and other health problems in a group of dairy herds in England', *Veterinary Record*, vol 139, pp486–490

Farm Animal Welfare Council (2009) 'Five Freedoms', available at www.fawc.org.uk/freedoms.htm, last accessed 11 February 2010

Harris, B. L. and Winkelman, A. M. (2000) 'Influence of North American Holstein genetics on dairy cattle performance in New Zealand', *Proceedings. New Zealand Large Herds Conference*, vol 6, pp122–136

NDFAS (2010) www.ndfas.org.uk, last accessed 24 May 2010

Pryce, J. E., Veerkamp, R. F., Thompson, R., Hill, W. G. and Simm, G. (1997) 'Genetic aspects of common health disorders and measures of fertility in Holstein Friesian dairy cattle', *Animal Science*, vol 65, pp353–360

Pryce, J. E., Veerkamp, R. F. and Simm, G. (1998) 'Expected correlated responses in health and fertility traits to selection on production in dairy cattle', *Proceedings of the 6th World Congress on Genetics Applied to Livestock Production*, Australia, vol 23, pp383–386

Smith, B. I., Kristula, M. A. and Martin D. (2007) 'Effects of frequent functional foot trimming on the incidence of lameness in lactating dairy cattle', *The Bovine Practitioner*, vol 41, pp137–145

Webster, A. J. F. (2009) 'The Virtuous Bicycle: A delivery vehicle for improved animal welfare', *Animal Welfare*, vol 18, pp141–148

Webster, J. (2005) *Animal Welfare: Limping Towards Eden*, Blackwell Publications, Oxford

Whay, H. R., Main, D. C. J., Green, L. E. and Webster, A. J. F. (2003) 'Assessment of the welfare of dairy cattle using animal-based measurements: Direct observations and investigation of farm records', *Veterinary Record*, vol 153, no 7, pp197–202

7

Cracking the Egg

Ian J. H. Duncan

Introduction

Ever since the publication of Ruth Harrison's book *Animal Machines* (Harrison, 1964), the question of the humaneness of battery cages for laying hens has been hotly debated. Harrison criticized the fact that birds were kept crowded together in a very unnatural environment, were denied access to fresh air and sunlight and were at high risk of 'cage layer fatigue', a disease that causes partial paralysis. In the UK, the publication of these facts led to such public outrage that the British Government formed a committee under the Chairmanship of Professor Rogers Brambell to investigate the whole topic of intensive animal production systems. They published their findings in 1965 in a report that is generally known as *The Brambell Report* (Command Paper 2836, 1965). This report was also very critical of battery cages. For example, the report stated:

> *Much of the ingrained behaviour is frustrated by caging. The normal reproductive pattern of mating, hatching and rearing young is prevented and the only reproductive urge permitted is laying. They cannot fly, scratch, perch or walk freely. Preening is difficult and dust-bathing impossible... The caged bird which is permitted only to fulfil the instinctive urges to eat and drink, to sleep, to lay and to communicate vocally with its fellows, would appear to be exposed to considerable frustration.* (Command Paper 2836, 1965, pp18–19)

However, the Brambell Committee's conclusions were not based on much hard evidence. In the subsequent 40 years, there has been a burgeoning of research into the welfare implications of cages and this paper summarizes the results.

A short history of battery cages

Animal production systems became ever more specialized in the first half of the 20th century, with the increasing demand for cheap food. Egg producers, who had previously kept relatively small flocks of hens outside with movable huts or inside on the floor in barns, had difficulty in scaling up their operations. The reasons for this were two-fold. These existing husbandry systems were not easy to automate and so had a large labour requirement that added considerably to the cost of egg production. The second reason was diseases transmitted through the gut. These included certain parasitic diseases such as coccidiosis (*Eimeria* spp.) and intestinal worm infestations as well as bacterial infections such as Pullorum disease (*Salmonella pullorum*), fowl typhoid (*Salmonella gallinarum*) and avian tuberculosis (*Mycobacterium tuberculosis avium*). These diseases were the bane of the poultry industry and the resultant 'fowl sick land' proved to be a formidable obstacle to scaling up.

Harrison (1964) describes cages being used for laying hens by Professor J. Halpin of the University of Wisconsin in 1911. Apparently these were even then in a 'battery' of three tiers. However, it is not clear whether the hens were being kept in this way to facilitate some research procedure or as a prototype for commercial production. It is known that cages for laying hens started to be produced commercially in the US and UK in the early 1930s and they started to catch on in the industry in the 1940s. The first cages were for single hens and still required a lot of labour. However, it quickly became apparent that there were lots of opportunities for automation. Feed and water could be delivered to the birds, manure could be removed from the house and eggs could be collected automatically at the touch of a switch.

There was a pause in the spread of the battery cage system during the Second World War, but in the 1950s it took off, driven by the demand for cheap food. Because birds in cages were separated from their faeces, the cycle of reinfection was broken, and many of the diseases that had held the egg production sector in check practically disappeared. At first, each cage held only one hen. However, it was quickly realised that economies could be made by having several hens in a cage. Today, five or six birds in a cage is normal practice.

Any egg farmer wishing to convert to battery cages was faced with a huge outlay in capital, but nevertheless the advantages of cages, including economic advantages, quickly became apparent, and the switch to cages continued apace. For example, the figures for the UK are shown in Table 7.1.

Ironically, most of this expansion occurred after Harrison (1964) and the Brambell Committee (Command Paper 2836, 1965) had sounded their warnings of the threat to welfare posed by battery cages.

Table 7.1 *Percentage of UK laying flock kept under the three main husbandry systems*

	Free range	Deep litter	Battery cages
1960	31	50	19
1980	2	2	96

Source: Sainsbury (1992)

How cages reduce welfare

Harrison (1964) and the Brambell Committee (Command Paper 2836, 1965) got it right in general terms: cages do reduce welfare. However, they missed some important and unexpected effects.

Frustration of nesting behaviour

I carried out one of the first investigations into the effects of battery cages on laying hen welfare and took as my starting point the statement made by the Brambell Committee, 'The caged bird which is permitted only to fulfil the instinctive urges to eat and drink, to sleep, to lay and to communicate vocally with its fellows, would appear to be exposed to considerable frustration' (Command Paper 2836, 1965, p19). I examined how hens reacted to a variety of frustrating situations in the laboratory, listed the symptoms shown and then looked for these symptoms in battery cages (Duncan, 1970). I showed that the biggest source of frustration is the lack of a nesting site in battery cages. Nesting behaviour in domestic fowl is a very interesting pattern shown by hens. It is not, as one might assume, triggered by the presence of a hard-shelled egg in the shell gland, but by hormonal events that take place at ovulation, 24 hours earlier (Wood-Gush and Gilbert, 1964, 1973, 1975). Once nesting behaviour is triggered, it is an extremely strong motivation. About one to one and a half hours before an egg is laid, the hen separates from the flock, gives a very characteristic pre-laying call and starts to seek a suitable nesting site. Once a secluded site is found, the hen enters and shows nest-building behaviour (Duncan and Kite, 1989). Most of this behaviour is prevented in a cage. The very early parts of the sequence look similar as the hen starts calling and seeking a nesting site. However, this soon develops into stereotyped back-and-forward pacing, the main symptom of severe frustration (Duncan, 1970; Wood-Gush, 1972). Most hens of light hybrid strains show these symptoms of frustration compared to a smaller (but still significant) proportion of birds of medium hybrid strains and this difference has a genetic basis (Mills et al, 1985a, 1985b). Moreover, it has been shown that hens are highly motivated to find a suitable nesting site and will work very hard to obtain one (Duncan and Kite, 1987). In fact it has been shown that hens will work as hard to gain access to a nesting site as they will to obtain food when they have been deprived of food for 28–30 hours (Follensbee et al, 1992). In my opinion, this thwarting of nesting behaviour that occurs for between one and one and a half hours on seven days out of eight throughout the laying year, has a huge negative impact on the welfare of laying hens.

Lack of social space

Like most social species, domestic fowl do not arrange themselves at random in the available space. There appear to be psychological forces that keep them apart; hens do not like to be crowded together. However, if given lots of room, they do not spread out as much as they can. There appear to be other psychological forces that pull them together; they are, after all, a flocking species. Using overhead cameras, Keeling and Duncan (1989) showed that hens kept at the usual cage space allowance try to space themselves out as much as possible suggesting that commercial stocking densities are far too tight. It has also been shown that decreasing the area per bird depresses egg production, reduces food consumption, lowers body weight and increases mortality (Hill and Binns, 1973; Wells, 1973). This almost certainly means decreased welfare at normal cage stocking densities.

The preceding discussion of social spacing behaviour oversimplifies the topic because domestic fowl space themselves differently depending on what they are doing. Keeling and Duncan (1991) observed a small population of domestic fowl living in a semi-natural environment. They found that birds were furthest apart when they were walking, closer when they were ground pecking, even closer when they were standing and closest when they were preening. Other studies have shown that domestic fowl will voluntarily come into contact with each other while roosting at night and while dust-bathing. The social spacing that hens prefer for walking, ground pecking and even standing, is not available to them at 'normal' cage densities. And of course, although hens in conventional cages may be able to space themselves appropriately for roosting and dust-bathing, these activities are denied them because perches and dust-baths are not provided.

Lack of physical space for certain postures and activities

It could be argued that dividing space up into 'social space' and 'physical space' is somewhat artificial, and that this section and the preceding one could have been rolled into one. However, scientists have often designed experiments thinking that they were dealing with the restrictive effects of either social space or physical space and their discussions of the topic mirror that belief. I have continued with the division but believe the two phenomena are closely linked.

It would seem obvious that the low height of battery cages would not allow hens to adopt the 'standing alert' posture that is very common in their repertoire, and this has, indeed, been shown to be true (Dawkins, 1985). In other experiments in which cage height and cage area were manipulated independently, some other less obvious effects were revealed. When height was reduced to that of battery cages, the incidence of head stretching, head scratching and body shaking were reduced, and, surprisingly, the length of time that hens spent sitting was also reduced. When cage area was reduced to that of battery cages, head scratching, body shaking and feather raising were performed at a lower rate and cage pecking at a higher rate (Nicol, 1987a). In another set of experiments, it was found that behaviour patterns that had been shown at a low incidence when space was restricted, showed a 'rebound' effect when the hens were given more space, suggesting that the spatial restriction had resulted in a build-up of motivation to perform these patterns (Nicol, 1987b). All of these results suggest that the dimensions of normal battery cages may compromise welfare by restricting the behavioural repertoire of the birds.

Another approach to investigating the effects on welfare of the restrictive space available in battery cages has been to 'ask' hens how much space they prefer. Both Hughes (1975) and Dawkins (1981) have shown that hens prefer more space than that available in conventional battery cages. The fact that the argument can be made 'the hen herself prefers more room' is even more powerful evidence that battery cages provide insufficient space.

Lack of roosting opportunities

The natural behaviour of both red jungle fowl and feral fowl is to roost in a perching posture high off the ground during the night. Domestic fowl kept in non-cage production systems also roost on high perches at night if these are provided (Blokhuis, 1984). Presumably, this behaviour has evolved to give protection from nocturnal predators. A battery cage both prevents hens from changing level at night and from adopting a normal perching posture, which is the natural posture for sleeping and resting (Blokhuis, 1984). However, hens do seem capable of adopting other postures for sleeping and resting, so do battery cages have an adverse effect on welfare through denying hens normal roosting? Olsson and Keeling (2000) investigated this question and found that hens, used to roosting on perches high off the ground, showed symptoms of frustration when they were denied access to the perches. In other experiments, they showed that hens would work quite hard, by pushing open a weighted door, in order to gain access to perches for roosting (Olsson and Keeling, 2002). This is strong evidence that battery cages reduce welfare by preventing hens from performing normal roosting behaviour.

Lack of foraging opportunities

It has been shown that jungle fowl (Dawkins, 1989) and feral domestic fowl (Savory et al, 1978) spend the major part of their day foraging for food. Although hens in battery cages have food continuously available to them, they cannot perform the normal motor patterns associated with foraging; that is, locomotion together with ground scratching and pecking, probing and flicking at items on the ground with the beak. In addition, feral domestic fowl consume leafy material that they tear from growing plants with their beaks (Savory et al, 1978). Most of these motor patterns are denied to birds in cages. They can peck at the food in the food trough and probe and flick at it with their beaks but the normal foraging sequence of scratching, stepping back and pecking at the scratched location is prevented by battery cages as is the tearing of leafy material. In addition, the body posture while performing feeding activities in the cage is very different from that assumed when foraging in an extensive environment. Moreover, the locomotion that normally accompanies foraging is prevented in cages. It would therefore seem that thwarting of foraging behaviour would be a possible source of reduced welfare in cages and would be a candidate for further investigation. It has been shown that foraging behaviour and the consumption of food may be driven by two distinct motivational systems and that the performance of foraging by hens may indeed be important for their welfare (Moffat and Duncan, 1999).

Lack of dust-bathing opportunities

During preening, chickens add oil to their feathers from a small gland in front of the tail called the uropygial gland. This serves to waterproof the feathers. However, this oil goes stale over time and chickens remove it regularly by means of dust-bathing. Approximately every two days, chickens kept under extensive conditions will find a location with dry dusty substrate, lie in it, peck at it and with vigorous wing movements incorporate the dust into their feathers. Any stale oil on the feathers adheres to the dust particles and is shaken out at the end of a dust-bathing bout. Obviously there is no dusty substrate within a battery cage and so hens are denied the functional consequences of dust-bathing – they cannot get rid of stale oil on their feathers by this means. Nevertheless, caged birds are seen to engage in dust-bathing behaviour occasionally. In ethological jargon this is described as a 'vacuum activity' – a behaviour pattern occurring in the absence of the stimuli that usually elicit it. The question then arises 'Are caged birds frustrated because they cannot dust-bathe properly?'

It turns out that dust-bathing behaviour is not motivated by a build-up of stale oil on the feathers but by external factors and in particular by the sight of a dusty substrate, high air temperature, presence of radiant heat, presence of radiant light and the sight and sounds of other birds dust-bathing (Petherick et al, 1995; Duncan et al, 1998). In the absence of the correct combination of external factors, dust-bathing behaviour will not be triggered. Generally this is what happens in battery cages. However, hens are often fed a dry dusty mash and occasionally the sight of this mash can trigger vacuum dust-bathing in cages. (Hens performing vacuum dust-bathing in cages are often seen with their heads over the food trough looking at the mash.)

So does the lack of dust-bathing opportunities lead to suffering in cages or not? Widowski and Duncan (2000) set out to answer this question and measure the importance of dust-bathing to hens by seeing if hens would work in order to gain access to a dust-bathing substrate. Hens were trained to push through a weighted swinging door to enter a goal box containing peat moss (their favoured dust-bathing substrate). The hens were tested to see how hard they would work when deprived and non-deprived of the opportunity to dust-bathe. The results suggested that there was tendency for deprived hens to work a little harder than non-deprived hens, but they were not clear-cut. These results are very difficult to explain using a 'needs' model of motivation in which deprivation leads to a state of suffering. They are much more consistent with an 'opportunity' model of motivation in which performance of the behaviour, when the opportunity presents itself, leads to a state of pleasure (Fraser and Duncan, 1998).

The conclusion is that although the lack of dust-bathing opportunities in cages may not lead to suffering, they may deprive hens of the opportunity to experience pleasure, and in assessing the overall quality of life that laying hens have, this may be important.

Lack of exercise

As has been discussed above, the lack of space and lack of stimulation in battery cages can lead to various frustrations in laying hens. However, there is an additional effect that can seriously impair their health and therefore their welfare. Being housed in battery cages means that laying hens do not get sufficient exercise to maintain bone strength and this manifests itself at two different times. Caged laying hens sometimes develop a paralysis around the time of peak production. This is known as 'cage layer fatigue' and is caused by fractures of the fourth and fifth thoracic vertebrae, which cause compression of the spinal cord (Riddell et al, 1968). It is almost never seen in hens kept in alternative husbandry systems, only in caged hens (hence its name). The fact that progression of the disease can be stemmed by transferring birds from cages to a floor system of management is further evidence that lack of exercise is a causative factor (Leeson et al, 1995). The fact that the disease appears at peak production is because this is the time of maximum demand for calcium for egg shells and bone reserves of calcium are depleted. A related phenomenon is the osteoporosis and bone weakness seen in hens at the end of a laying year (Leeson et al, 1995). Modern laying hens can lay 320 eggs in a year, which requires lots of calcium for egg shells. Each day, bone reserves of calcium are used for the next egg's shell and then replenished from dietary sources. However, there is a gradual calcium loss from the skeleton over the course of a year and this is exacerbated by lack of exercise. The amount of exercise required to strengthen bones may be very little. It has been shown that the addition of a perch to a cage can increase bone strength (Abrahamsson and Tauson, 1993). At the end of a laying year hens from battery cages have less lower limb bone strength than hens from non-cage systems (Knowles and Broom, 1990; Nørgaard-Nielsen, 1990). It is not known whether this condition is painful in itself, but even if it is not, the hen has still a dreadful price to pay. Many hens have bones broken as they are removed from battery cages at the end of a laying year. In a British survey, 29 per cent of hens from battery cages were found to have broken bones just before they were stunned at the slaughter house and most of this damage occurred as the hens were being removed from the cages (Gregory and Wilkins, 1989). The fact that battery cages are not very well designed for the removal of birds, and at the end of a laying year hens are worth very little and so are handled with little care, contribute to the problem.

Do battery cages have anything to offer?

Hygiene

As was pointed out earlier in this chapter, the hygienic advantages of cages were very obvious when they were first introduced. Many of these advantages remain today, but the disease problems can now often be managed in other ways. For example, there are now blood tests for most of the salmonella diseases and with good management they can be eliminated. Intestinal parasites including coccidia (*Eimeria* spp.) may prove to be a more difficult problem to get rid of. However, it appears that effective vaccines for coccidia are on the horizon and they will certainly be a great benefit.

Until then, intestinal parasite control will depend on very high management standards. Of course one possible strategy would be to house hens on slatted or wire floors to maintain the separation of birds and droppings. I would argue strongly against this solution, because of the lack of foraging and dust-bathing opportunities that are part of these systems. So any indoor husbandry system requires litter management of the highest standard. The litter must never be allowed to become cold and wet or these parasitic diseases will inevitably appear. Outdoor or free range husbandry systems must employ strict rotation of the pasture or foraging areas so that there is no build-up of pathogenic organisms. Finally strains of hen should be sought that have some resistance to the various parasites. A more severe problem may actually come from external parasites such as lice, fleas and various species of mite and in particular Red Mite (*Dermanyssus gallinae*). The incidence of these parasites is generally higher in non-cage systems than in cages and is increasing (Chauve, 1998).

Group size

Battery cages commonly contain four to six hens. This is much closer to the basic social unit of 10–20 birds seen in jungle fowl (Collias et al, 1966) and feral domestic fowl (McBride et al, 1969), than that occurring in alternative husbandry systems, which commonly have many hundreds and sometimes thousands of hens in the one group (Blokhuis and Metz, 1992). Of course, domestic fowl may be able to adapt to life in large groups and so their welfare may not be adversely affected by extensive husbandry systems. What do the birds themselves prefer? Early preference tests showed that hens prefer smaller rather than larger groups and familiar hens over strangers (Hughes, 1977; Dawkins, 1982). However, these tests used only very small groups (up to six birds for the large groups). Later preference tests have shown that, when available space is properly controlled, most hens prefer a group size of 70 birds to one of four birds (Lindberg and Nicol, 1993, 1996). Of course this group size is still very much smaller than the very large groups of several hundred or several thousand hens in non-cage husbandry systems. Nevertheless, it opens the possibility of encouraging the formation of subgroups within the main population by using environmental features and 'furniture' to divide the total area visually (but not physically) into smaller precincts in which subgroups of 50–100 hens would feel comfortable.

Non-cage husbandry systems for laying hens

There are several alternative husbandry systems to battery cages available for laying hens. These have been gradually increasing in popularity in Europe as the general public becomes aware of the welfare problems associated with cages and as governments take action to ban cages. For example, conventional battery cages will be prohibited in the European Union from January 2012 (European Commission, 1999). The proportion of laying hens housed in non-cage systems in the EU increased from 8.7 per cent in 1995 (Agra CEAS, 2006) to 32.2 per cent in 2008 (European Commission, 2009). However, this trend is not occurring on a world-wide basis as shown in Table 7.2. It is worrying that the ease of operation and economic advantages of battery cages seem to be inducing developing countries to switch to them without the welfare implications being properly considered.

Non-cage husbandry systems include free range operations in which the hens can spend much of their time outside. The winters in northern Europe and much of north America prevent free-range from being used throughout the year. However, it can be used seasonally. 'Barn systems' are indoor husbandry systems with many variations (see for example Appleby et al, 1992).

Most of the variations are based on the traditional 'deep litter' system but with various structures included to make more use of the height of the barn. By this means, more hens can be kept per unit area making the barn more economical and, in addition, the barn is kept warmer from the metabolic heat of the birds. The husbandry system is usually named after the structures that are added, and so there are Percheries, Aviaries and various 'Tiered Systems'. A further variation is to have a roofed area next to the barn enclosed by mesh so that the hens have access to the outdoors without

Table 7.2 *Egg production by housing system in six of the top egg-producing countries (accounting for 64.5% of eggs worldwide)*

Country	% Cage	% Barn	% Free range
China	65	10	25
US	98	0	2
India	78	22	0
Japan	95	0	5
Mexico	100	0	0
Brazil	100	0	0

Source: World Poultry (2008)

it being true free range. Occasionally, a barn system with such a 'Winter-garten' attached will be used to house the hens during the winter and they will be given access to free range during the summer.

Also available are various types of 'furnished cage',[1] which are larger and, have more headroom than standard battery cages. They also include a secluded nesting area and perches and sometimes a dust-bath. For a more detailed description see Appleby et al, 2002.

The welfare advantages and disadvantages of these systems will now be discussed briefly.

Free range

The welfare advantages of free range systems are fairly obvious. The hens have almost complete behavioural freedom. They can choose a nesting site and show full nesting behaviour. Of all husbandry systems, free range offers the best possibilities for foraging. Perching and roosting behaviour are allowed. Moreover, the hens can space themselves according to the activity they are engaging in and they can choose an appropriate microclimate. An additional welfare advantage is that at the end of a laying year the hens' skel-etons are much stronger and they are therefore at much lower risk of bone breakage during catching and transportation.

The welfare disadvantages are that the hens are exposed to weather extremes and also to the risk of predation. There is also a risk of infection by internal parasites particularly *Coccidia* spp. and a risk of infestation by external parasites especially Red Mite (*Dermanyssus gallinae*). Strangely, the risk of feather pecking and cannibalism is also greater in all non-cage systems. This is probably because this behavioural abnormality can be culturally transmitted. Therefore, if there are a few primary peckers in a population of hens, their feather pecking behaviour can spread widely if they are in a very large group but it is greatly restricted when they are divided up a few hens to a cage.

Barn systems

Many of the welfare advantages of free range are also present for the various barn systems. Hens can choose nesting sites and show nesting behaviour. They have more room than in battery cages, but high densities are often used in barn systems and so the hens may not be able to space themselves exactly as they want. Usually foraging behaviour will be available to hens in barn systems, but not always. In an attempt to improve hygiene and air quality some barn systems are completely slatted with no loose substrate for foraging. Considering that hens spend 60–70 per cent of their active time foraging (Savory et al, 1978), an all-slatted system would seem to deny the birds a very important activity. An all-slatted system also prevents the hens from dust-bathing. It should be said, however, that most barn systems *do* have substrate available for foraging and dust-bathing. The majority of barn systems make use of the height dimension and have raised perches available so that birds can show normal resting and roosting behaviour. At the end of a laying year, bones of hens from barn systems are much stronger than those of hens from battery cages (Knowles and Broom, 1990; Nørgaard-Nielsen, 1990) meaning that they are at much lower risk of suffering bone breakage during catching and transportation. However, this should be balanced against the fact that hens in barn systems have been shown to suffer from a higher incidence of broken bones during the laying year. Nicol et al (2006) found that 60 per cent of hens in aviaries had old fractures (mainly of the sternum) at the end of the laying year, presumably sustained when they bumped into things in the barn. Barn systems also protect hens from extreme weather and from predators.

The welfare disadvantages of barn systems are that the presence of a loose substrate, essential for foraging, means that ammonia and dust levels are often high (Appleby et al, 1988) and this can lead to respiratory problems. The group size in barn systems is often several thousands of hens, much larger than the preferred group size of 50–100 hens (Lindberg and Nicol, 1993, 1996). However, there is also evidence that hens in very large groups (3000–16,000) can be very calm and non-aggressive (Whay et al, 2007). The risk of feather pecking and cannibalism is higher than in cages. There is also a small risk of infection with internal parasites but this is lower than in birds on free range. A bigger problem is the risk of Red Mite (*Dermanyssus gallinae*) infestation. Not only does this parasite pose a welfare risk through the direct effect of blood loss but it is also being implicated as a vector of other diseases (De Luna et al, 2008). Currently there is a big international effort to seek a solution to this problem (Mul et al, 2009).

Furnished cages

The welfare advantages of furnished cages are that they combine the few good points of conventional cages, namely hygiene and small group size, with the ability of the hens to show nesting behaviour, perching and sometimes dust-bathing behaviour. However, the perching opportunity, although allowing the hen to adopt the normal resting posture, may not give it the height above floor level that seems to be a desired component of perching (Olsson and Keeling, 2000). Furnished cages also provide more space than battery cages, but the birds are still very crowded and this is certainly a welfare disadvantage. In addition, they do not allow normal foraging behaviour, which, as previously stated, is an important activity for hens.

Conclusions

The welfare advantages for non-cage husbandry systems for laying hens are overwhelming. These include allowing hens to engage in nesting, roosting and foraging behaviour all of which are important to hens. The extra space available in non-cage systems also allows hens to adopt normal postures and carry out various comfort behaviour patterns that are constrained by cages. In addition, hens in non-cage systems can space themselves socially according to their behavioural activity, which caged hens cannot do. Dust-bathing is also a possibility for non-caged hens. Although prevention of dust-bathing may not cause suffering, there is now good evidence that its performance may generate pleasure.

Another major advantage of non-cage husbandry systems is that hens are able to move vigorously, flap their wings, change level by jumping and so on. This exercise gives them very good protection against cage layer fatigue at peak production and fragile skeletons at the end of a laying year. They are therefore at much lower risk of suffering painful bone breakages.

The few welfare advantages that cages appear to have are increased hygiene and small group size. However, a high standard of management with non-cage husbandry systems can generally counteract this apparent advantage. The one area that requires further improvement is the control of external parasites in non-cage systems.

Note

1   Also known as 'modified' or 'enriched' cages.

References

Abrahamsson, P. and Tauson, R. (1993) 'Effect of perches at different positions in conventional cages for laying hens of two different strains', *Acta Agriculturæ Scandinavica, Section A, Animal Science,* vol 43, pp228–235

Agra CEAS Consulting (2006) 'Trends in laying hen numbers and the production and consumption of eggs from caged and non-caged production systems', final report for Eurogroup for Animal Welfare, Agra CEAS Consulting, Wye, UK

Appleby, M. C., Hogarth, G. S., Anderson, J. A., Hughes, B. O. and Whittemore, C. T. (1988) 'Performance of a deep litter system for egg production', *British Poultry Science,* vol 29, pp735–751

Appleby, M. C., Hughes, B. O. and Elson, H. A. (1992) *Poultry Production Systems: Behaviour, Management and Welfare*, CABI, Wallingford

Appleby, M. C., Walker, A. W., Nicol, C. J., Lindberg, A. C., Freire, R., Hughes, B. O. and Elson, H. A. (2002) 'The development of furnished cages for laying hens', *British Poultry Science*, vol 43, pp489–500

Blokhuis, H. J. (1984) 'Rest in poultry', *Applied Animal Behaviour Science*, vol 12, pp289–303

Blokhuis, H. J. and Metz, J. H. M. (1992) 'Integration of animal welfare into housing systems for laying hens', *Netherlands Journal of Agricultural Science*, vol 40, pp327–337

Chauve, C. (1998) 'The poultry red mite *Dermanyssus gallinae* (de Geer, 1778): Current situation and future prospects for control', *Veterinary Parisitology*, vol 79, pp239–245

Collias, N. E., Collias, E. C., Hunsaker, D. and Minning, L. (1966) 'Locality fixation, mobility and social organization within an unconfined population of red jungle fowl', *Animal Behaviour*, vol 14, pp550–559

Command Paper 2836 (1965) *Report of the Technical Committee to Enquire into the Welfare of Animals kept under Intensive Livestock Husbandry Systems*, Her Majesty's Stationery Office, London

Dawkins, M. S. (1981) 'Priorities in the cage size and flooring preferences of domestic hens', *British Poultry Science*, vol 22, pp255–263

Dawkins, M. S. (1982) 'Elusive concept of preferred group size in domestic hens', *Applied Animal Ethology*, vol 8, pp365–375

Dawkins, M. S. (1985) 'Cage height preference and use in battery-kept hens', *Veterinary Record*, vol 116, pp345–347

Dawkins, M. S. (1989) 'Time budgets in red jungle fowl as a basis for the assessment of welfare in domestic fowl', *Applied Animal Behaviour Science*, vol 24, pp77–80

De Luna, C. J., Arkle, S., Harrington, D., George, D. R., Guy, J. H. and Sparagano, O. A. E. (2008) 'The poultry Red Mite *Dermanyssus gallinae* as a potential carrier of vector-borne diseases', *Annals of the New York Academy of Sciences*, vol 1149, pp255–258

Duncan, I. J. H. (1970) 'Frustration in the fowl', in Freeman, B. M. and Gordon, R. F. (eds) *Aspects of Poultry Behaviour*, British Poultry Science, Edinburgh, pp15–31

Duncan, I. J. H. and Kite, V. G. (1987) 'Some investigations into motivation in the domestic fowl', *Applied Animal Behaviour Science*, vol 18, pp387–388

Duncan, I. J. H. and Kite, V. G. (1989) 'Nest site selection and nest building behaviour in domestic fowl', *Animal Behaviour*, vol 37, pp215–231

Duncan, I. J. H., Widowski, T. M., Malleau, A. E., Lindberg, A. C. and Petherick, J. C. (1998) 'External factors and causation of dustbathing in domestic hens', *Behavioural Processes*, vol 43, pp219–228

European Commission (1999) 'Council Directive 1999/74/EC laying down minimum standards for the protection of laying hens', *Official Journal of the European Communities*, L203, pp53–57, 3.8.1999, available at http://eur-lex.europa.eu/LexUriServ/LexUriServ.do?uri=OJ:L:1999:203:0053:0057:EN:PDF, last accessed 25 May 2010

European Commission (2009) 'Laying hens by type of housing', CIRCA (Communication and Information Resource Centre Adminstrator), available at http://circa.europa.eu/Public/irc/agri/pig/library?l=/poultry_public_domain/public_statistics&vm=detailed&sb=Title, last accessed 12 February 2010

Follensbee, M. E., Duncan, I. J. H. and Widowski, T. M. (1992) 'Quantifying nesting motivation of domestic hens', *Journal of Animal Science*, vol 70 (Supplement 1), p50

Fraser, D. and Duncan, I. J. H. (1998) ' 'Pleasures', 'pains' and animal welfare: Toward a natural history of affect', *Animal Welfare*, vol 7, pp383–396

Gregory, N. G. and Wilkins, L. J. (1989) 'Broken bones in domestic fowl: Handling and processing damage in end-of-lay battery hens', *British Poultry Science*, vol 30, pp555–562

Harrison, R. (1964) *Animal Machines*, Vincent Stuart, London

Hill, A. T. and Binns, M. R. (1973) 'Effect of varying population density and size on laying performance', Proceedings of the 4th European Poultry Conference, London, pp605–609

Hughes, B. O. (1975) 'Spatial preference in the domestic hen', *British Veterinary Journal*, vol 131, pp560–564

Hughes, B. O. (1977) 'Selection of group size by individual laying hens', *British Poultry Science*, vol 18, pp9–18

Keeling, L. J. and Duncan, I. J. H. (1989) 'Inter-individual distances and orientation in laying hens housed in groups of three in two different-sized enclosures', *Applied Animal Behaviour Science*, vol 24, pp325–342

Keeling, L. J. and Duncan, I. J. H. (1991) 'Social spacing in domestic fowl under semi-natural conditions: The effect of behavioural activity and activity transitions', *Applied Animal Behaviour Science*, vol 32, pp205–217

Knowles, T. G. and Broom, D. M. (1990) 'Limb bone strength and movement in laying hens from different housing systems', *Veterinary Record*, vol 126, pp354–356

Leeson, S., Diaz, G. J. and Summers, J. D. (1995) *Poultry Metabolic Disorders and Mycotoxins*, University Books, Guelph, Canada

Lindberg, A. C. and Nicol, C. J. (1993) 'Group size preferences in laying hens', in Savory, C. J. and Hughes, B. O. (eds) 'Proceedings of the Fourth European Symposium on Poultry Welfare', Universities' Federation for Animal Welfare, Potters Bar, pp249–250

Lindberg, A. C. and Nicol, C. J. (1996) 'Space and density effects on group size preferences in laying hens', *British Poultry Science*, vol 37, pp709–721

McBride, G., Parer, I. P. and Foenander, F. (1969) 'The social organization and behaviour of the feral domestic fowl', *Animal Behaviour Monographs*, vol 2, pp125–181

Mills, A. D., Wood-Gush, D. G. M. and Hughes, B. O. (1985a) 'Genetic analysis of strain differences in pre-laying behaviour in battery cages', *British Poultry Science*, vol 26, pp182–197

Mills, A. D., Duncan, I. J. H., Slee, G. S. and Clark, J. S. B. (1985b) 'Heart rate and laying behaviour in two strains of domestic chicken', *Physiology and Behaviour*, vol 35, pp145–147

Moffat, L. A. and Duncan, I. J. H. (1999) 'Effects of food and substrate deprivation on foraging behaviour in laying hens', Proceedings of the Annual Meeting of the Canadian Society of Animal Science, Charlottetown, Prince Edward Island, 7–11 August 1999, *Canadian Journal of Animal Science*, vol 78, p586

Mul, M., Van Niekerk, T., Chirico, J., Maurer, V., Kilpinen, O., Sparagano, O., Thind, B., Zoons, J., Moore, D., Bell, B., Gjevre, A.-G. and Chauve, C. (2009) 'Control methods for *Dermanyssus gallinae* in systems for laying hens: Results of an international seminar', *World's Poultry Science Journal*, vol 65, pp589–599

Nicol, C. J. (1987a) 'Effect of cage height and area on the behaviour of hens housed in battery cages', *British Poultry Science*, vol 28, pp327–335

Nicol, C. J. (1987b) 'Behavioural responses of laying hens following a period of spatial restriction', *Animal Behaviour*, vol 35, pp1709–1719

Nicol, C. J., Brown, S. N., Glen, E., Pope, S. J., Short, F. J., Warriss, P. D., Zimmerman, P. H. and Wilkins, L. J. (2006) 'Effects of stocking density, flock size and management on the welfare of laying hens in single-tier aviaries', *British Poultry Science*, vol 47, pp135–146

Nørgaard-Nielsen, G. (1990) 'Bone strength of laying hens kept in an alternative housing system compared with hens in cages and on deep litter', *British Poultry Science*, vol 31, pp81–89

Olsson, I. A. S. and Keeling, L. J. (2000) 'Night-time roosting in laying hens and the effect of thwarting access to perches', *Applied Animal Behaviour Science*, vol 68, pp243–256

Olsson, I. A. S. and Keeling, L. J. (2002) 'The push-door for measuring motivation in hens: Laying hens are motivated to perch at night', *Animal Welfare*, vol 11, pp11–19

Petherick, J. C., Seawright, E., Waddington, D., Duncan, I. J. H. and Murphy, L. B. (1995) 'The role of perception in the causation of dustbathing behaviour in domestic fowl', *Animal Behaviour*, vol 49, pp1521–1530

Riddell, C., Helmboldt, C. F., Singson, E. P. and Matterson, A. D. (1968) 'Bone pathology of birds affected with cage layer fatigue', *Avian Diseases*, vol 12, pp285–297

Sainsbury, D. (1992) *Poultry Health and Management* (3rd ed.), Blackwell Scientific Publications, Oxford

Savory, C. J., Wood-Gush, D. G. M. and Duncan, I. J. H. (1978) 'Feeding behaviour in a population of domestic fowls in the wild', *Applied Animal Ethology*, vol 4, pp13–27

Wells, R. G. (1973) 'Stocking density and colony size for caged layers', Proceedings of the Fourth European Poultry Conference, London, pp617–622

Whay, H. R., Main, D. C. J., Green, L. E., Heaven, G., Howell, H., Morgan, M., Pearson, A. and Webster, A. J. F. (2007) 'Assessment of the behaviour and welfare of laying hens on free-range units', *Veterinary Record*, vol 161, pp119–128

Widowski, T. M. and Duncan, I. J. H. (2000) 'Working for a dustbath: Are hens increasing pleasure rather than reducing suffering?', *Applied Animal Behaviour Science*, vol 68, pp39–53

Wood-Gush, D. G. M. (1972) 'Strain differences in response to sub-optimal stimuli in the fowl', *Animal Behaviour*, vol 20, pp72–76

Wood-Gush, D. G. M. and Gilbert, A. B. (1964) 'The control of the nesting behaviour of the domestic hen. II. The role of the ovary', *Animal Behaviour*, vol 12, pp451–453

Wood-Gush, D. G. M. and Gilbert, A. B. (1973) 'Some hormones involved in the nesting behaviour of hens', *Animal Behaviour*, vol 21, pp98–103

Wood-Gush, D. G. M. and Gilbert, A. B. (1975) 'The physiological basis of a behaviour pattern in the domestic hen', *Symposia of the Zoological Society of London*, vol 35, pp261–276

World Poultry (2008) 'Egg production by housing system in selected countries', *World Poultry*, vol 24, pp20–21

8

Cheap as Chicken

Andy Butterworth

Chickens are the most common farmed animal on solid earth. In 2010, around 45 billion chickens will be reared for meat. This means that in the year 2010, about the same number of chickens will be reared for meat as some people estimate the number of humans who have ever lived on the planet – ever. How has it come to be that chicken has taken such a global hold, consuming vast amounts of cereal and protein, and, in turn, being consumed in vast numbers? In this chapter I discuss chickens reared exclusively for meat. Chicken has become a barometer for least cost production and can be purchased in many supermarkets for less money by weight than some fruits and vegetables. With this quite astounding mass of animals in the 'care' of mankind, is it inevitable that the global chicken business will keep as its central philosophy – 'cheap as chicken'?

Chicken history

The history of chicken intensification builds on the importance of three key ingredients – the move of rural people to towns and cities, electricity – enabling increased poultry house size and use of fossil fuel to enable the harvesting and transport of feed from previously unachievable distance. But first, the conventional history... In Asia today you will see backyard birds that are identical in nature to the first domesticated chicken, the red jungle fowl – light, agile birds, still able to fly for short distances and to climb trees, rear young and produce eggs and (a little) meat.

The Ancient Egyptians appear to have reared poultry in large groups and to have developed technology for incubating large numbers of eggs. But the first birds with a lineage to the current birds reared on a commercial scale were imported from Asia to the USA and Europe and can still be traced through existing lines such as the Java chicken. However, most chickens were reared in backyards and small flocks until the 1930s where expanding city populations in the US created a demand for fresh poultry meat and around cities like New York and Chicago, farms started to rear increasingly large numbers of birds in large outdoor flocks and housed in small group coops.

In the 1930s most farms did not consider the use of electricity for provision of services to animal houses and so the use of automated systems for ventilation were rare. Farmers at this time were aware of the effects of poor air and house environment on their birds, and by the late 1930s farmers were beginning to experiment with keeping birds in more 'controlled environments'. By 1940 many birds were still given access to the outdoors but the 'range houses' that they used at night were becoming increasingly large and complex, with feeding systems, lighting and ventilation systems starting to develop as electricity became more widely used. A quantum change in philosophy became apparent in the 1950s with the development of the first widely used fully housed poultry systems – and chicken rearing started to move entirely into houses. To enable this, systems to provide food automatically, to control ventilation and to regulate house temperature were developed and by 1960, the house design used throughout the world was established, and would be quite recognizable to many poultry farmers today. As the birds moved indoors, people began to forget poultry as regular 'farmyard' animals and started to see poultry as a low cost food available in the food market, rather than at the farmer's gate.

Today, globally over 75 per cent of poultry meat now comes from birds reared entirely indoors. Of the 25 per cent not reared under controlled conditions, the majority are farmed at family or local farm or subsistence level (particularly in Africa and Asia) – however, a small percentage of birds are beginning to 'return to the paddock' as consumers in the developed world choose to purchase free range or organic poultry.

Chicken houses

If so many thousands of millions animals are kept indoors, are these houses good places for them to live out their lives? Although there are variations on what constitutes a modern chicken house there are some generally recognizable features. Birds are kept on a floor covered in a layer of litter. This litter can be wood shavings, rice hulls (the husk of the rice grain), chopped straw, peat or chopped newspaper. Around the world, wood shavings and rice hulls are the dominant material. The floor under the litter is concrete in many countries, but in some places the floor is earth. The birds are kept under a low roof. The roof might be supported by solid walls, or by mesh or netting walls in tropical countries. In some countries, particularly in Asia, some broiler houses are now 'double deckers' with two levels under one roof. Some houses are made from very modern materials – steel and concrete, with steel or aluminium sheet clad walls and roof, whilst others, even some very large houses, particularly in South America, are made from local timber, with clay tiles on the roof.

Light is usually provided by electric lights, and in fully enclosed houses, light levels are generally quite low – at around 20–40 lux, just a bit more than the minimum you would need to read this book. In houses with mesh sides, or with pop holes to the outdoors, light levels can be very high. One reason commonly given for keeping birds at low light levels has been that this keeps them 'calmer' and so less likely to become excitable and to 'pile up' and suffocate. The vast differences in light levels seen in different poultry houses around the world make this argument a bit hard to maintain.

In most large poultry houses artificial ventilation is required. Ventilation brings in fresh air and removes ammonia and dust – and in countries with high temperatures, ventilation is combined with cooling in the form of evaporative cooling and tunnel ventilation to reduce house temperatures. Almost all current ventilation systems rely on electricity, and in some countries electrical energy is one of the highest unit costs for poultry production. Without electricity, the houses used to produce most of the world's chicken could not function.

At the beginning of life for many chicks, particularly in cooler northern or southern countries, the poultry house is heated using gas, oil or electrical heaters to provide the warmth for the young chicks. As the chicks grow they produce heat (the litter also produces heat as it breaks down) and so quite quickly the 'heating' can usually be turned off, and needs to be replaced by ventilation – even in cold countries.

To feed the birds, two main types of feeders are used, pan feeders, which look like bowls and are supplied with feed by screw augers that carry the feed from silos outside the house. In the other (less common now) system, a metal U-shaped track contains a chain and as the chain is pulled around the track it carries food from a silo outside the house. In the smaller poultry houses seen in some countries, or in free range mobile houses, the feed can be provided for the birds in 'hopper feeders' – small hoppers of feed that may need to be filled from sacks by hand.

Water is provided to the birds via pipes with 'nipples' or 'nipples and cups' (or less commonly) via 'bell drinkers' – all of these systems allow centralized supply of water and drinking points throughout the house. The water system is also used to supply vaccinations and sometimes (less commonly) soluble antibiotics, to the birds via the water.

The control systems for the temperature, ventilation, light, feed and water are usually centralized and controlled by electronic sensors that measure the humidity and temperature in the house. Sample groups of birds are weighed so that the farmer can see how they are growing in comparison with 'expected performance' and sometimes feeding patterns or lighting patterns are adjusted to alter the rate of growth. Today the birds in many houses can be found at a density of up to 20 birds for every metre square (birds m²). In broiler production language, this is usually converted to 'kg per metre squared' (kg/m²) – removing the reference to the animal as an individual, and turning the discussion into one about how much 'biological mass in kg' a certain space can carry. Some studies have shown that not only is the space available a factor, but that the way in which the farmer cares for the animals is critically important (Jones et al, 2005).

The whole chicken farm is often part of a vertically integrated business where the chicks and the feed are supplied by the company and the grown birds are taken to the company slaughterhouse. Around the world, variations on this 'recipe' for intensive poultry production exist – some birds are kept in mixed sex groups (as they were hatched) whilst some are kept in single sex groups after having been sexed at the hatchery. If birds have been separated into single sex groups, then they will usually be fed under different 'programmes' to match the different rates of growth of males and females.

Chicken litter

Chickens in almost all intensive systems live their lives on a bed of litter. In well-managed litter, birds can rest without becoming dirty and they can care for their feathers by preening. If the litter is dry and deep, birds can, and will, dust-bathe, and can forage and scratch – although there is nothing for them to eat in litter. Poorly managed litter can make the birds cold, wet and dirty, can prevent them from cleaning themselves and preening, make dust-bathing impossible and contact with the litter can produce skin lesions on the feet, hocks and breast. Litter can become wet and sticky, or caked with a hard surface and even oily or greasy. Conversely, litter can also become very dusty – making the house atmosphere damaging to the birds' lungs and respiratory system. Litter 'quality' is at the heart of a chicken's life experience – and, unfortunately, litter quality is also often poor.

Poultry farmers are fully aware of the importance of litter to their birds – but they must juggle the temperature and ventilation in the house, the supply of water via the drinkers, the gut health of the birds and the labour required to manually manage litter. If the farmer drops any of these multiple juggling balls, then litter quality can change very fast – and the life of the birds can become uncomfortable, wet, sticky, dirty and, if skin lesions become established, pain-filled.

Chicken – a 'low margin per head' business

As the intensive poultry house has developed, two major trends have emerged. The first trend is towards large farms, managed by only a handful of people. The second trend is toward a 'low margin per bird' but 'lots of birds' philosophy.

Some farms in northern Europe and the US may have 250,000 birds and be managed by two people. In fact, the man to bird ratio can fall to almost unbelievable levels – the grass is cut by contractors, the birds are taken from the house for slaughter by visiting catching teams, the houses are cleaned by a cleaning gang at the end of the growing 'cycle' and the houses and machinery are repaired by outside technicians when required. The farmer walks the houses looking for sick birds, keeps an eye on the monitoring system and plans the feed deliveries and the schedule for slaughter and restocking the houses. The farmer in fact becomes a skilled technician with vast numbers of animals under his or her technical care. Does this matter? If the automated machinery can ventilate and feed the birds, then is there an issue? When a person can farm so many animals, can it be realistic to expect that sick or disadvantaged animals will receive a sensible degree of 'care'? I call this dilemma the 'margin for care problem'.

Some animals, for example some pigs and cows, are kept with care and compassion and even with affection, and then killed and eaten – but some animals are kept with routinely only the smallest 'margin for care' – they are considered as economic 'particles', whose needs are met within the constraints of economic capacity. I suggest that a fundamental concept of animal care in farming systems could, and should, be the concept of 'margin for care'.

What is the margin for care in chicken farming?

1 Making care a virtue: if society only portrays farmers as efficient and shrewd businessmen and technicians, but fails to portray them as people who are 'permitted', or expected and 'enabled to care' for farmed animals, then farmers may lose sight of the fact that at the heart of their business is the care of animals – creation of places and systems where animals find a place of care to live out their time on earth.

Can farmers be given permission by society and by the farming community to recognize the need for care? If so, how can we (society) reward farmers who show a real ability to 'care' for animals rather than an ability to make systems work and to make (small) profit margins from each animal?

2   An economic margin for care: farming is a business and some farming businesses operate on the tightest of economic margins – if a lone farmer can be given authority to care, with technical assistance, for 250,000 animals, should it be required that the business can show sufficient economic margin to provide the basics of an animal's life – light, space, thermal and physical comfort and low levels of fearfulness or disease? If there is almost no economic or conceptual margin, then the care capacity may be pared back to a skeleton. If there is no margin, then there is little or no capacity to invest in improvements in animals' lives, or to create environments that do more than provide the absolute basics.

If a business that looks after so many sentient animals cannot show that it can provide acceptable, or even high, levels of these requirements, then is the farm, and the farming system, showing contempt for the concept of 'margin for care' – in other words, is it operating on a basis and a platform based on unsustainably low levels of animal concern?

The idea that 'care' is at the very core of farming seems alien to some people – some may say that farming is a serious business. Farmers might say 'we don't have time for these cosy concepts, the farming world is a hard cold place, these are farm animals not pets, and we need to provide only what is necessary to achieve production goals'. But is that right? It may be that this philosophy (farming as a tough economic activity) erodes a very simple basic principle – that if we permit animal-derived food to become a commodity alone, we shift the concept of margin to its economic meaning only and we fail to ask or even to expect that farming provides sufficient 'margin for care'. Without a concept of chickens (or other farmed animals) as worthy of investment of care, countless animals across the world will live their lives (and each life is important to each animal) with only the thinnest separation from the bottom of the care barrel.

Animals farmed in a minimal margin system may 'produce' – they may weigh sufficient at the time of slaughter, they may grow as expected – but that may be all we can say. We probably aren't able to say that we (as society) showed high levels of care – we provided only as much care as the low levels of margin permitted.

Chicken business

The chicken business in 2010 is global. Although many companies rear chickens in their own country, the birds are all part of a global bird family linked by a small number of broiler chicken genotypes – the selected lines of chicken genetics which provide the great-grandparent, grandparent, breeder birds and finally the eggs that will hatch out to be the birds that are reared for meat.

From the 1930s to the 1950s, cross-breeding of different lines of poultry derived from Cornish and White Plymouth Rock birds resulted dramatic changes in the characteristics of the bird. In the 1950s almost all chicken was sold as 'whole birds', for roasting or 'broiling', and the whole production system was named after this – the broiler industry. Today, in 2010, and in contrast to the 1950s, whole bird sales now form only a small percentage (between 10 and 30 per cent depending on country) of the poultry meat market. The majority of chicken meat is now sold as fillets, breast portions and cuts, or as a part of processed dishes. Some restaurant chains like KFC and Nandos have built their whole brand on chicken as an easy, and sometimes fast, food. Well, chickens are fast – fast-growing. In the 1950s broilers birds took about 70 days to reach a body weight of about 1.5kg. In 2010, a broiler bird takes about 35 days to reach about 2.2kg and can put on up to 100g of weight a day toward the end of the production period. This is a phenomenal rate of growth – most birds on the supermarket shelf are less than 40 days old!

It is not even that simple – because there is a part of the chicken that has actually developed its growth rate faster than the rest of the bird. The breast muscle is the most valuable part of the bird, and the yield of breast meat has increased much beyond that expected by overall bird growth. The yield (a measure based on percentage of the bird's weight) of the pectoralis major, the breast muscle, has doubled in the last 30 years from 8.3 per cent (1977) to 15.4 per cent (2005), whilst the yield of meat in the leg has not changed much since 1957 (7.5 per cent) compared to 8.2 per cent in 2005 (Canadian Poultry, 2009).

Chicken production races

Feed is the single most expensive input into poultry production and so, not surprisingly, the rate at which feed can be converted into fowl has its own vocabulary in poultry production language. The 'performance' of a chicken producer can be assessed using a production efficiency factor (PEF), and feed conversion efficiency (FCE) or feed conversion ratio (FCR) are measures used to describe (and to compare) how efficiently chicken can turn feed into flesh. Relatively high feed conversion is sometimes used as an argument for chicken as an efficient world feeder – converting low grade cereal and plant-based protein and energy into concentrated animal protein at a ratio of 1kg chicken for about 2.6kg feed – but what does this actually mean?

If you give chickens 2.6kg of feed containing about 20 per cent protein, you will get about 1 kg of chicken meat. Only fish have a higher 'feed conversion rate', as ruminants usually require about 7 or 8kg of feed for each kg of beef or lamb produced. However, although the chicken 'conversion deal' sounds OK, there are a number of catches.

- Catch 1: dry feed is turned into 'wet' chicken, in other words the chicken growing from the feed is made up largely from water (as all living animals are) but the feed fed is dry, so the actual 'conversion ratio' of feed to chicken meat is much higher than first apparent.
- Catch 2: chicken feed is the final product of a long chain; as the cereals, soya, vegetable proteins, fats and fish meal are harvested, collected and processed, there are losses at every stage, so the true 'conversion ratio' – the amount of carbohydrate and protein in total required to rear a chicken – is much greater than that at first made apparent in feed conversion ratios.
- Catch 3: the broiler is not made up entirely of high value 'breast meat', but legs, bones, intestines and all the parts which make up a whole animal. Because breast meat provides the highest economic returns, there is profound pressure to increase breast muscle 'yield' in disproportion to the other parts of the bird, but there are also parts of the birds that are not well used as food. This means that the actual final 'true' conversion rate for the edible parts of the bird can be much lower, at around 1kg for 4.5kg of feed.

The practical impact of Catch 3 is that the broiler breeders strive to create high breast muscle mass, potentially to the detriment of other characteristics. The balancing act that the poultry geneticists engage in is now very complex. They juggle fertility characteristics (hatchability, liveability of chicks) with skeletal shape, growth rate, muscle conformation of the different parts of the animal (yield), nutritional 'needs', disease resistance characteristics and overall survival to production age. What has been created over numerous generations is a highly mouldable farm creature, the characteristics of which can be altered within only a few generations to suit changes in the market.

One, rarely considered, outcome of this selection is the loss of flight. The most common farmed animal is a bird – which can't fly! A race of non-flying birds with hyper-developed flight musculature but nowhere to fly to and no capacity to fly anyway! Of course, chickens are not alone in being functionally flightless – but chickens (and farmed turkeys) have been very selectively created this way, thus tethering them to the farm.

In a 2010 report on broilers, the European Food Safety Authority confirmed some of the major welfare problems of broilers and the impact of breeding and feeding practices, saying: 'the major welfare concerns for broilers are leg problems, contact dermatitis, especially footpad dermatitis, ascites and sudden death syndrome. These concerns have been exacerbated by genetic selection for fast growth and increased food conversions' (EFSA, 2010).

Figure 8.1 Broiler breeder birds: Mixed males and females in a house with nest boxes

Breeder birds

An important part of the meat chicken system is the parent stock – the birds that produce the eggs that will then hatch and be placed on the production farms. These birds are kept on farms that look more like laying hen farms, with both male and female birds (in a ratio of about 1:10) and with nest boxes for collecting the fertilized eggs. These birds are kept at high levels of biosecurity, but in general must be kept at variable levels of feed restriction so that the males do not become too big, heavy or lame, and so that fertility levels are kept 'high'. The degree of feed restriction can be quite severe. In some countries 'skip a day' feeding is permitted, where the male birds are only fed every other day, or on a rotation of feeding days. The levels of frustration and hunger induced by restricted feeding are difficult to 'measure', but when feed is provided, it is clear from the behaviour of the birds that they are highly motivated to feed!

In a court case brought by Compassion in World Farming in 2003, the judge agreed that such birds were in a state of 'chronic hunger' – while dismissing the case (R (Compassion in World Farming Limited) v Secretary of State for Environment, Food and Rural Affairs).

Chicken journeys

The global poultry business has some strange aspects. The chickens themselves are closely related – originating in the elite stock of the breeder companies in countries including Scotland, the US and Australia – and the relatives of these birds are spread by truck and aeroplane across the whole globe. They find themselves finally, after another generation, in broiler breeder sites where they grow (but more slowly than their productive children) and mate to produce the fertile eggs that hatch to make the broiler chickens that will be eaten in huge numbers.

The eggs produced by the breeder birds are handled on a grand scale in hatcheries. For every chicken eaten (45 billion a year) more than one egg is required, because a percentage do not make it to the farm. The chicks pip their way into the world fuelled only on yolk and their inbuilt genetic knowledge. Everything they do is self-learned – no teenage chicken siblings to transfer knowledge, no adult birds passing on feeding, nesting and living tips. The chicks are a single age group. This results in their own (perhaps instinctive) decisions and reactions arising without any input from previous generations. You may think that this is, well … obvious – but I think this needs some reconsideration. Amongst farmed animals, only fish abandon their young (and not all of them do this) – pigs, cattle, sheep, goats – and chickens if given the option – nurture and *teach* their young. But in the automated, grand scale production of chickens, this natural and quite fundamental generation step is completely side stepped. Broiler chickens receive all of their 'being a broiler' education from their peers, and from the inherent behaviours and physiological skills that they carry deep in their DNA.

Perhaps it should not be forgotten that one aspect of the centralized intensive farming of chickens is the complete severing of this transmission of 'learned knowledge' from animal generation to generation. Who knows what the long-term effects of this may be? Perhaps these animals will slowly lose the 'inherent DNA knowledge' on which we rely at present to give them the 'out of the shell' skills to feed, sleep and grow for human gain. Perhaps by severing this link we eventually create animals so dependent on human support that they can no longer be self supporting. Perhaps we have already? Some people would say that some strains of broiler chicken are so hyper-adapted that they cannot in fact survive in outdoor systems, or to maturity, without feed control and restriction. This might be a warning alarm.

Another surprising aspect of global chicken production is the fact that cargo ships are criss-crossing the world's oceans carrying (a) feed stuffs for chickens and (b) chicken meat. This sounds quite expected, but it is not. In Asia, leg meat is widely eaten and valued, whilst in the developed world, the price differential between leg meat and white breast meat is very large. So…the developed world produces leg meat and exports it to other parts of the world, whilst the countries that will import leg meat, export breast meat to the developed world. As well as this 'crossing' of trade, the feed required by chickens also makes some surprising journeys: soya from Brazil, Argentina, China and India crosses with ships carrying cereals from Russia, Canada, the US and Australia. These cross with boats carrying poultry products to Asia from Europe, from Europe to Asia and from South America to Australia, Europe and Eurasia.

The criss-crossing of feed and chicken product make complex spider web patterns across the globe and the carbon cost, food miles and economic implications of this meshwork become almost impossible to unravel. Global decisions based on cost and availability determine whether cereals, soya and fish protein are sold to people or to feed animals. The routes that the big boats carrying these foods steer dictate fat or famine for many people across the globe. The chicken in your pre-prepared dish is a relative of a great-grandparent bird grown in Australia from Scottish ancestors. The chicken may have eaten soya from Brazil, cereal from the Ukraine and been transported to you around the Pacific, and then by road from a factory in Holland where it was packaged.

Compare that travelogue with the story of your chicken in 1928 – you went to the butcher and bought a whole, but quite expensive bird, grown by a farmer 20 km away using local corn. How can it be sensible to do all this cross transporting? It can, and clearly does, make 'economic sense' whilst:

- fuel is cheap;
- chicken is a commodity that can be 'grown' in almost any country;
- feed mills search for least cost and maximized efficiency feed formulations;
- the carbon consumptive and globally unsustainable networks of poultry trade are considered 'the norm'.

If any of these things change, the spider web of the poultry trade will get thinner until it cannot hold up and the links between consumer and chicken will probably become more direct again.

Chicken ills

Chickens can suffer disease, like any other animal. The difference is that, if disease enters a poultry house containing 25,000 birds then it is *not* possible to hospitalize or to examine and treat individuals, so chickens are always considered as a mass, rather than as individuals. Chickens are routinely protected from some diseases, for example Infectious Bursal Disease Virus, by vaccination at the hatchery, or via the drinking water on farm. They may also be protected from some parasitic diseases by vaccination and until recently (and still the case in some countries) by using anti-parasiticidal drugs acting against a gut parasite, coccidia.

Broilers can also suffer a range of 'production' diseases, including lameness (sometimes called leg weakness), skin lesions on the feet (pododermatitis) and on the back of the scaly part of the leg (hock burn), skin damage due to poor litter (breast blister and cellulites) and also respiratory and heart disease affected by diet, genetic susceptibility and by house air and litter quality. Of these, lameness is probably the most widely seen disorder (Broom and Reefmann, 2005). The prevalence in broilers from moderate to severe lameness (scores of three to five in a lameness scoring scale) close to the time of slaughter has been reported to range up to 27.3 per cent (Knowles et al, 2008). In other words over a quarter of chickens are in pain for several days before slaughter.

In general, the philosophy in intensive indoor-reared chicken is to protect by vaccination and to make farms into fortresses that block access to most visitors – and so hopefully to uninvited pathogens and disease. This fortress farm philosophy may protect the animals from disease, but it makes the poultry business almost invisible. Houses containing 25,000 animals may have no windows and be indistinguishable from commercial storage sheds. This, in my opinion, raises a simple philosophical dilemma; as the poultry business has disappeared inside, the mental and 'actual' contact between chickens and consumers has disappeared. There is no longer a clear view of what a 'chicken farm is' or even of how a farmed chicken looks and behaves.

Stocking densities can reach high levels toward the end of production when the birds are at maximum weight

Figure 8.2 Broiler (meat) chickens

This invisibility is in direct contrast to the actual dominance of chicken in the food market. This is clearly not an intentioned hiding away, but it may finally have disturbing implications, with so many animals and such a dominant arm of farming effectively 'hidden'.

Chicken needs

Linked to the issue of invisibility is another question that concerns some people, but which does not occur to many. Is it right and just that the most common farmed animals on the planet very often *never*, in their entire lives, see the sun that warms the houses where they live, and powers the plants that produce the cereal and protein that they eat? Some people may consider this a fundamental aspect of a real life – to be illuminated by the sun, whilst others would consider this an irrelevant or even subversive question – bringing a widely used farming system into question. Many scientists would question whether sunlight *per se* is of any importance at all. If an animal can see, if enough artificial light is provided, then why is there an issue? Some studies have shown that birds can see with different spectral sensitivity to mammals and that sunlight may be more essential to them than previously thought (Bennett, 2007).

Most farmers would say that they provide all that is required for the bird's life, including adequate light. I still think that keeping the most numerous farmed animal on the planet mostly hidden from the sun is an almost 'spiritual' issue that should be considered in the debate on what we, as mankind, do to the animals we use for our purposes.

In the same vein, it is clear that chickens are provided with the anatomy and physiology required for a life outdoors. They have feathers that can provide waterproofing, thermal cover and the ability to fly (a little), they can forage and dust-bathe and nest and fight. Some birds are asked to use all of their physiological adaptations and we see that free range and backyard birds shrug off rain, they bathe in dust and puddles, react to wind and calm and sun and cold by choosing where they lie, sit, perch, roost, fight and sleep. For the majority of the enormous population of chickens farmed, these adaptations are only just skimmed and the birds are provided with a highly moderated environment and protected from these real challenges, but also never asked to use the systems that they brought with them from the egg.

A practical farmer might say, 'yes … isn't it great, we have made their lives so much easier, they don't need to worry about birds of prey or foxes, they are protected from disease, food is only a step or two away, and the day is pretty much free from any unexpected challenges. The birds can mostly eat and *grow*.'

An ethologist may say – 'here's a list of things I know that chickens can (and do) do when given a chance – scratching, foraging, dust-bathing, climbing, perching, taking shade, seeking heat, avoiding wet etc... and I've just been watching the birds in the poultry house, and mostly they seem to do little but kick back and *grow*, whilst those outdoors are exercising their physiological capabilities'.

A third person may come along and ask – 'all I want to know is – how many survive, and how much do they weigh, how do they *grow*?' – and for both systems (indoor reared and free range) figures can be provided. And the last enquirer might say – 'well, it's clear, these indoor birds really can *grow*, and fewer die because we can slaughter them earlier because they have grown so well!'

And a fourth person may say – 'well this growing thing is OK, but what about the *life* of these birds – if we give them a choice by opening the doors of the house out into paddocks, what do the chickens choose? We find that some chickens choose to spend a lot of their time outside and some don't, and that on warm dry days many do, and on cold wet days, many don't.'

So we have some complex decisions – do we (as a society of chicken eaters) think that growth rate and 'cheap chicken' are so overwhelmingly important that we forget choice, body conformation and the ability to express physiological capacities?

What do chickens actually need? Nobody can truly say, but we can probably hazard some pretty good guesses. Chickens are sentient, capable of feeling pain and avoiding sources of stress and distress when they are give a choice. Free range and backyard birds show a large range of adaptive behaviours that are not 'engaged' in intensive indoor conditions. Society asks a lot of the chicken and it expects that there will be affordability, availability and practical invisibility. The chicken is expected to perform at phenomenal levels – from egg to plate in 40 days – and to return something to the farmer in the way of 'margin' but to be at the level of 'readily affordable' for most consumers.

These are the 'average human' expectations of the chicken. What the chicken actually needs is probably quite different, and applying human expectations is of course likely to attract criticisms of 'anthropomorphism', but it is probably fair to consider that a chicken, like almost any animal, will seeks safety, some degree of comfort, both physical and thermal, and avoidance of fear, bullying, pain and disease (when possible – disease is a part of many animals' lives, including humans), space to exercise the physiological capabilities that it has been provided with and company. Food, water, light, air that is not damaging to breathe and a surface to live on that does not cause skin lesions and that enables some behaviours that chickens commonly perform: dust-bathing, scratching, pecking and foraging. If the systems that we create for the lives of this enormous number of animals do not meet these quite uncomplicated criteria, then it is possible that we are permitting, in fact promoting, a quite extraordinary failure – the chickens are providing what we expect, and we are not returning a large part of their 'needs'.

If we as consumers wish to make a difference through our purchases, we can. We can identify chicken reared to specific standards (Free range, Organic, Freedom Food, *Label Rouge*, Devonshire Red etc.) or which come from systems of production that we perceive as being likely to offer better living conditions for the birds. We might choose to purchase only 'fresh meat' if we are concerned about transportation of poultry meat, as fresh poultry is unlikely to have been transported over long distances. We can also make choices on the content of processed food containing chicken – where does it come from and what was the system of production? Moves are starting to be made in assurance systems and labelling to provide information not only on the type of farm where the animal was kept, but also information on the welfare status of the animal from these farms (animal outcome-based assessment), and although this type of information is not yet common on poultry products, in the future this may allow the discriminating consumer to make more informed ethical purchasing choices for chicken. Finally, we could avoid chicken that is 'cheap as chicken'. As long as consumers search for poultry meat at very low cost, then production systems that create low margins per bird and low individual 'animal value' will remain dominant.

Summary

- Meat chickens come from a family of closely related 'genotypes';
- the global chicken business relies on a network of feed supplies – with complex interlinked effects of transport and local effects in the places where the feeds are grown;
- there is a huge amount of cross-country and cross-continent transport of feed and of refrigerated poultry product;
- the need for high levels of biosecurity, and the types of closed housing systems used, has lead to 'fortress farming' and so poultry production has now become almost 'invisible' despite its massive scale;
- birds live in an environment completely controlled by man, reliant on electrical ventilation and the quality of their lives can be hugely impacted by litter, air and feed and water quality and the technical influence of the farmer;
- systems of production, although similar in many parts of the world, do make big differences to the life experience of the birds, and it is possible to provide consumers with information not only on the system of production – but also the welfare status of the birds from different farms;

- using system and welfare information, we can make ethical purchasing decisions that influence a shift away from 'cheap as chicken';
- in 2010, around 45 billion chickens will be reared for meat. If the systems that we create for the lives of this enormous numbers of animals do not meet quite uncomplicated criteria for animal 'comfort, care and welfare', then it is possible that we are permitting, in fact promoting, a quite extraordinary failure – the chickens are providing what we expect but we are not returning a large part of their 'needs'.

References

Bennett, A. T. D. (2007) 'Avian color vision and coloration: Multidisciplinary evolutionary biology', *The American Naturalist*, vol 169, Supplement 1

Broom, D. M. and Reefmann, N. (2005) 'Chicken welfare as indicated by lesions on carcases in supermarkets', *British Poultry Science*, vol 46, issue 4, pp407–414

Canadian Poultry (2009) www.canadianpoultrymag.com/content/view/953/38/, last accessed 23 March 2009

Compassion in World Farming (2007) 'Supermarkets and farm animal welfare: Raising the standard', available at www.ciwf.org.uk/resources/publications/food_industry/default.aspx, last accessed 24 December 2009

EFSA (European Food Safety Authority) (2010) 'Scientific opinion on the influence of genetic parameters on the welfare and the resistance to stress of commercial broilers', EFSA, Parma, adopted 24 June 2010

Jones, T. A., Donnelly, C. A. and Dawkins, M. S. (2005) 'Environmental and management factors affecting the welfare of chickens on commercial farms in the UK and Denmark stocked at five densities', *Poultry Science*, vol 84, pp1155–1165

Knowles, T. G., Kestin, S. C., Haslam, S. M., Brown, S. N., Green, L. E., Butterworth, A., Pope, S. J., Pfeiffer, D. and Nicol, C. J. (2008) 'Leg disorders in broiler chicken: Prevalence, risk factors and prevention', *PLoS ONE*, vol 3, issue 2, e1545, doi:10.1371/journal.pone.0001545

R (Compassion in World Farming Limited) v. Secretary of State for Environment, Food and Rural Affairs (2004) EWCA Civ 1009

Scientific Committee on Animal Health and Animal Welfare (SCAHAW) (2000) 'The welfare of chickens kept for meat production (Broilers)', European Commission, Health and Consumer Protection Directorate-General, March 2000, available at http://ec.europa.eu/food/fs/sc/scah/out39_en.pdf, last accessed 24 December 2009

9

Sustainable Pig Production: Finding Solutions and Making Choices

Alistair Lawrence and Alistair Stott

The positioning of this chapter next to the one entitled 'Cheap as Chicken' is deliberate. There are many similarities between the production of pigs and broiler chickens for meat. First, the scale of these production industries has grown enormously since the Second World War. FAO statistics calculate that the global number of pigs rose from 400 million in 1961 to 900 million in 2007 (FAOSTAT, 2009). China dwarfs the rest of the world in having approximately 500 million pigs compared to, for example, 150 million pigs in the entire enlarged EU. This is matched of course with consumption patterns and pig meat continues to grow as a highly popular form of animal protein.

Second, there has been a significant and similar intensification process applied to both broiler chickens and pigs in terms of:

1 the size of farms, the numbers of animals per farm, and the stocking density (numbers of animals per unit space) on farms;
2 the development of specialised housing to increase efficiency including significant reductions in the labour required to remove manure and to feed and water the stock;
3 the application of animal science (genetics, nutrition, reproductive biology) to further increase the efficiency of production.

Third, this growth in global production in pigs and poultry and the drive to make the farm-end of the supply of pig meat more efficient has brought with it a range of concerns over the long-term sustainability and ethical standards of pig production. In pigs the first such animal welfare concerns to be raised were in Ruth Harrison's book *Animal Machines* (1964). Harrison focused mainly on the use of close confinement systems to house sows and the heavily stocked and 'barren' environments used for growing pigs. It is worth just briefly painting a picture of the intensive pig farm that Harrison described.

By the 1960s, pig farmers had largely adopted a system for housing pregnant sows in narrow crates that allowed them to stand and lie but not to turn round; the sows generally defecated and urinated on to a 'slatted floor' (literally a floor with long vertical spaces through which the manure or 'slurry' could escape) and were fed and watered from a trough at their head. They were prevented from 'escape' either by being enclosed by the crate itself or by a 'tether' chain that restrained them around the neck. Obviously this type of system largely prevented the 'natural behaviour' of sows (D'Eath and Turner, 2009), such as rooting in the soil and exploring their environment. In addition, close confinement systems for pregnant sows were also accused of causing specific and severe problems. For example, pregnant sows in crates often performed a type of abnormal behaviour known as stereotypic behaviour (repetitive and persistent behaviour patterns, which are often also seen in confined zoo animals, and are generally seen as indicating poor welfare) and they also often developed painful physical problems such as leg and joint weakness and urinary tract infections. A similar system had also been developed to house sows before and after giving birth (known as farrowing in pigs). Farrowing crates were very similar in design to those used for pregnant sows but had an additional area around the crate to allow the piglets to move around the sow and locate the udder. Farrowing crates were also heavily restrictive of sow behaviour, and this was a particular issue at farrowing because before they give birth sows are strongly motivated to build a nest (in the forest environment, where pigs evolved, this would be from branches and other materials the sow could gather and carry). The justification for the use of farrowing crates (over and above increasing stocking density and being easier to manage) was that without them the sow was more likely to crush and kill her piglets.

Once weaned from their piglets, sows would be served by the boar and returned to the pregnancy crate. The piglets would be placed into systems designed to provide these young animals with extra hygiene and heat, and also with specialized feeding and watering equipment. After this 'weaner' stage the piglets would be moved on to 'grower' accommodation. Although there were a wide range of systems used, they typically used 'slatted floors', solid walls and other than providing food and water were generally devoid of other features. Harrison's book emphasized the 'barrenness' of these environments, which did not provide the growing pigs the opportunities to explore and develop as they would have in a more natural setting. In addition these systems involved high stocking densities, often with poor light and air quality and proneness for pigs to develop physical problems from being constantly on solid floors (e.g. lameness and leg swellings). Although the natural weaning age of piglets is at around 17 weeks, in intensive farms piglets are usually removed from the sow at three to four weeks of age.

Astonishingly, not a lot has changed since the publication of Harrison's book (e.g European Food Standards Agency, 2007). The UK, to be followed by the EU in 2013, has banned the practice of closely confining pregnant sows in crates (although farrowing crates remain unaffected by this legislation) and other less confining systems (e.g. keeping sows in outdoor units) have become more popular. However, globally, pregnant sows are still very commonly housed in crates during pregnancy. The farrowing crate is also widely used and very little has changed with respect to the housing of weaned and growing pigs. In addition since the issue of pig welfare was first raised by Harrison other pig welfare issues have been raised including:

- the weaning of piglets at an early age in order to more quickly encourage the sow back into oestrous;
- the 'clipping' (using clippers to remove part) of the tusks and tails of the young pig: the clipping of tusks to prevent the piglets damaging each other and the sow when struggling to get access to her teats for milk, and the clipping of tails to help control the occurrence of tail biting where pigs bite and damage each others' tails;
- the mixing of pigs into new (non-litter) groups that often causes fighting between unfamiliar animals;
- the practice of food restricting pregnant sows that has been associated with the development of stereotypic behaviours (especially when sows are housed in crates during pregnancy);
- and as with other farm species: the issues relating to the transport and slaughter of pigs including pigs' responses to transport and slaughter;
- the wider effects on health and welfare of selective breeding for production traits (Rauw et al, 1998).

In addition to these welfare concerns, pig production is similar to all forms of livestock farming in being increasingly under review in terms of its wider environmental sustainability (Kanis et al, 2005). Many of these concerns mirror those for other species and perhaps particularly broiler breeders. Issues of particular concern with pigs are:

- the use of land or food stuffs to feed pigs that could be used to directly feed humans and questions relating to the overall efficiency of converting plant to animal protein;
- the 'carbon footprint' of pig production taking account of all the relevant inputs and outputs;
- the very large and concentrated amount of animal waste produced and associated issues of how to use or destroy this waste with minimum harm to the environment including to water sources;

- the wider environmental impacts of producing food stuffs for pigs in terms of environmental destruction, area of land involved, use of chemicals and impacts on the soil;
- the potential for harm to humans from zoonotic diseases passing between pigs and humans and from practices (e.g. the use of antibiotics) used to control diseases on densely populated pig farms.

What has further complicated the arguments over the sustainability of livestock farming, including pig production, has been the re-emergence of concerns over 'food security' and the very real likelihood that there will be food shortages in the future (Cabinet Office, 2008). The complication here is that the need to feed a growing global population combined with changes in affluence, which are increasing global demand for meat, could counter the aims of making livestock production more sustainable, not least because of the wider environmental effects of an increase in animal numbers if demand for animal products is unrestrained.

In this short chapter we will look at ways in which science can help balance (optimize) the various demands for increased sustainability that are likely to affect pig production in the medium term. We will also touch on wider policy areas that will need attention if sustainability goals are to be met in the face of a growing demand for pig meat.

An obvious place to look for solutions that could help improve the sustainability of pig production is at farm level. The majority of the welfare issues about modern pig production that concern the public are on-farm and many of the wider environmental issues also are farm-based (e.g. the majority of the carbon footprint of livestock production occurs before the animals (or products) leave the farm).

At the simplest level we could look at optimizing 'components' of systems to improve sustainability. Piglet mortality is an example of a system-component that remains high at close to 20 per cent including stillbirths; it is both an important welfare issue (involving much suffering), but also a waste of resources in terms of the environment and the farm business. Recent research has been looking at the potential to improve piglet survival through genetic selection for survival traits. In one study, sows from lines of pigs that showed higher survival were compared with a control population, farrowing on an outdoor unit in Scotland (Welfare Quality® Fact Sheet, 2009a). The results were encouraging with the high survival lines showing improved piglet survival over two subsequent generations. Important points raised by this work are that:

- The lines of pigs that were compared had been bred in intensive conditions yet still showed improved survival in a very different (outdoor) environment suggesting that the same characteristics support high survival in both indoor and outdoor conditions. One implication of this is that there should be no need to set up a separate outdoor breeding programme aimed at increasing piglet survival.

- One of the traits the high survival sows showed was greater care when lying hence lessening the risk of crushing their piglets (a major cause of piglet deaths in systems that do not use the farrowing crate to protect piglets).
- The findings represent an example where it is possible to align a number of sustainability goals. The use of genetic selection to reduce piglet mortality in a non-crate system is a significant welfare gain and could help in the future to reduce the reliance on the farrowing crate (see also below). At the same time, increased piglet survival also reduces the environmental impact of pig farming (by reducing waste) and also contributes positively to the profitability of the farm business.

Of course not all examples are so clear-cut in terms of their sustainability gains. A common problem in pig production is the fighting that occurs between unfamiliar pigs when they are mixed. This fighting causes injury and stress to the pigs and results in reduced weight gain, again representing a production and environmental waste. Research has demonstrated the potential of selecting genetically against the aggressive personality types that are at the basis of this problem (Welfare Quality® Fact Sheet, 2009b). There appear to be few if any negative welfare implications of such a selection, with likely benefits in terms of reducing the amount of fighting and wider impacts (e.g. loss of weight gain) that occurs when pigs are mixed. However, as this approach involves genetic selection to reduce what is effectively a 'natural' (if potentially damaging behaviour) in pigs, it may raise ethical concerns about breeding for pigs to 'fit the system' as opposed to using a 'system that fits the pigs'.

These sorts of concerns are even more apparent in the issue of tail biting. Tail biting (where pigs nibble and then bite each others' tails, often causing substantial injuries), is similar to a number of welfare problems in confined animals, including feather pecking in hens and fin nibbling in farmed fish, in that it occurs unpredictably and, once triggered, can escalate very rapidly through a pen, group or tank. There is no single factor underlying tail biting but the behaviour is generally only found in the barren conditions seen in intensive conditions, probably as a result of a number of factors (Bracke, 2008) coming together to trigger the initial exploration and biting of tails. As a result there is no easy solution to this problem and the practice of tail docking is still widely used to prevent the occurrence of tail biting. There is evidence of a genetic basis to the behaviour; however, the use of genetics to resolve a problem that appears to directly reflect keeping pigs under suboptimal conditions seems even more questionable than in the case of aggressive behaviour at mixing. There are examples of systems where tail biting is a rarity and tail docking is not required and one way forward could be system-level analysis to explore how to optimize conditions to better fit the pigs' needs (and hence reduce the risk of tail biting), taking account of other constraints including financial ones.

An example of this sort of approach is 'PigSafe', a project aimed at resolving the farrowing crate dilemma by developing an indoor pen system that would allow the sow freedom to behave whilst at the same time protecting the piglets from being crushed by the sow. In this project, economic model-ling is being used to optimize the 'interests' of the sow (e.g. to nest build), the piglet (e.g. to be safe) and the farmer (e.g. to make a business profit). Although at an early stage, the analysis suggests that using current research knowledge there is in fact only a relatively small difference in physical and financial performance between crates and a so-called 'designed-pen' which provides for improved sow and piglet welfare (see Figure 9.1).

The 'PigSafe' example illustrates the benefit of using a systematic analysis to help us understand how to optimize systems to meet different sustainability goals, in this case welfare and financial performance.

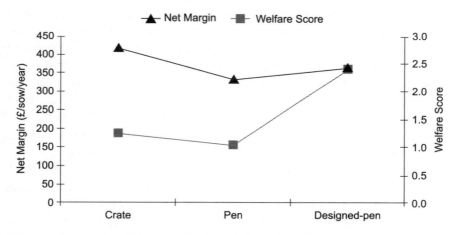

Note: See also www.sac.ac.uk/research/projects/theme/animalwelfare/. The welfare score was developed from an extensive review of the welfare impacts of different components and systems on sow and piglet welfare. The difference between a Pen and Designed-Pen is that in the latter case there are built-in features to protect the piglet from being crushed. The results suggest that a designed-pen can deliver substantial improvements to welfare at a relatively small cost. Further work is exploring how this 'gap' in financial performance could be further reduced.

Source: www.sac.ac.uk/research/

Figure 9.1 Preliminary results from the 'PigSafe' economic modelling illustrated as the 'trade-off' between the financial and welfare 'performance' of different farrowing systems

One study (Stern et al, 2005) looking at sustainable pig production at the whole farm level, developed future scenarios for pig farming that fulfilled different sustainability goals (animal welfare, low environmental impact and product quality and safety). This study found that no one scenario fulfilled all sustainability goals: animal welfare was the most economically costly, the economic scenario had potential welfare costs (the pigs being housed indoors to control environmental impacts) and the product-quality scenario had the highest environmental impact. The study represents one way of considering how we design more sustainable pig production systems; future work could explore in more detail how we optimize across future scenarios in order to meet different sustainability targets.

Another study (Ashworth et al, 2009) that has examined the wider impact of improving pig welfare on other sustainability goals, used an economic approach known as partial equilibrium (PE) modelling. PE modelling can be used to understand how a 'perturbation' (in this case an improvement to animal welfare) 'ripples' through to affect the wider environment and economy. Here PE was used to explore how improving animal welfare could affect both trade in pig meat and environmental pollution including greenhouse gas (GHG) emissions. Using a relatively straightforward welfare issue (the use of high fibre diets fed to sows in pregnancy to improve piglet survival) the PE modelling found, associated with a reduction in piglet mortality, an improvement in trade volumes and a small environmental benefit; in other words the analysis was able to quantify positive effects on multiple sustainability goals (a 'win-win-win' scenario).

There is clearly much work still to be done in evaluating how variants on pig systems can affect a range of sustainability goals, and in many ways the studies that are reported here are just starting points for more detailed analyses to come. The work on system 'components' such as using genetic selection to reduce aggression in pigs, has tended to be conducted by biologists and often there has been no analysis of the wider impacts of such technical approaches. On the other hand the more recent 'systems-level' work needs to be expanded and developed to deal with the 'real-life' complexity of pig production systems. Nonetheless there is clearly much promise in using economic and other integrative approaches, to understand the wider impacts of technical innovations in pig production, and of the changing directions of pig production systems to meet specific sustainability targets such as substantial reductions in emissions.

All the foregoing examples have involved the 'supply end' of the problem, looking at how farms can become more sustainable and the 'trade-offs' between achieving different sustainability goals. It is of course equally important to consider the 'demand side' of the sustainability issue. Recently there has been a growing interest in how we could translate animal welfare standards into a labelling scheme that would allow consumers to make more informed choices (Lang and Heasman, 2004) and there are increasing examples of 'carbon' labelling on animal products. However, labelling animal products to encourage consumers to purchase more sustainable products has a number of potential deficiencies not least because it may only affect a relatively small segment of the market. That is why some (Lang and Heasman, 2004) have suggested that we will need more than voluntary shifts in consumer behaviour if we are to seriously address the question of 'what can the planet afford for us to eat'. Other policy levers that have been proposed to change our consumption patterns include fiscal measures (e.g. a tax on animal products) or even some form of food rationing. It is important to begin to analyse the potential impact of such approaches not just in terms of their impact on consumer behaviour, but also on the production base that will still be required to produce animal products into the foreseeable future.

In conclusion, pig production systems have developed in a rather piece-meal fashion, largely driven by short-term financial considerations of how to produce more pig meat at least cost. The future will see increasing pressure on pig production to become more aligned with other sustainability goals, not least because it will be required to play its part in reducing GHG emissions and in improving animal welfare. The approaches described in this chapter that use integrative modelling to provide a systematic analysis of options, will be increasingly required to provide the 'road map' to achieving these longer-term sustainability targets. The success of this integrative approach will be largely dependent on there being a much closer alignment of biologists, modellers and social scientists than in the past. The benefits should be a transparent set of options for policy makers, farmers and retailers on how to best tackle the pressing problems wrapped up in the production of pig meat for our consumption.

References

Ashworth, C. J., Toma, L. M. and Hunter, M. G. (2009) 'Nutritional effects on oocyte and embryo development in mammals: Implications for reproductive efficiency and environmental sustainability', *Philosophical Transactions of the Royal Society B*, vol 364, pp3351–3361

Bracke, M. B. M. (2008) 'RichPig: A semantic model to assess enrichment materials for pigs', *Animal Welfare*, vol 17, pp289–304

Cabinet Office (2008) 'Food matters: Towards a strategy for the 21st century', Strategy Unit, Cabinet Office, UK

D'Eath, R. B. and Turner, S. P. (2009) 'The natural behaviour of the pig', in Marchant-Forde, J. N. (ed.) *The Welfare of the Pig*, Springer Science and Business Media B.V., The Netherlands

European Food Standards Agency (2007) 'Animal health and welfare aspects of different housing and husbandry systems for adult breeding boars, pregnant, farrowing sows and unweaned piglets – Scientific Opinion of the Panel on Animal Health and Welfare', *EFSA Journal*, vol 572, pp1–13

FAOSTAT (2009) http://faostat.fao.org/site/339/default.aspx, last accessed 10 December 2009

Harrison, R. (1964) *Animal Machines*, Vincent Stuart, London

Kanis, E., De Greef, K. H., Hiemstra, A. and van Arendonk, J. A. M. (2005) 'Breeding for societally important traits in pigs', *Journal of Animal Science*, vol 83, pp948–957

Lang, T. and Heasman, M. (2004) *Food Wars: The Battle for Mouths, Minds and Markets*, Earthscan, London

Rauw, W. M., Kanis, E., Noordhuizen-Stassen, E. N. and Grommers, F. J. (1998) 'Undesirable side effects of selection for high production efficiency in farm animals: A review', *Livestock Production Science*, vol 56, pp15–33

Stern, S., Sonesson, U., Gunnarsson, S., Oborn, I., Kumm, K. I. and Nybrant, T. (2005) 'Sustainable development of food production: A case study on scenarios for pig production', *Ambio*, vol 34, pp402–407

Welfare Quality® Fact Sheet (2009a) 'Improving piglet survival', available at www.welfarequality.net/everyone/41858/5/0/22, last accessed 10 December 2009

Welfare Quality® Fact Sheet (2009b) 'Reducing aggression in pigs through selective breeding', available at www.welfarequality.net/everyone/41858/5/0/22, last accessed 10 December 2009

Part 3

The Implications of Meat Production for Human Health

10

Industrial Animal Agriculture's Role in the Emergence and Spread of Disease

Michael Greger

The first major period of disease since the beginning of human evolution probably started approximately 10,000 years ago with the domestication of farm animals (Armelagos et al, 1996). Human measles, for example, which has killed roughly 200 million people over the last 150 years, probably arose from a rinderpest-like virus of sheep and goats (Weiss, 2001). Smallpox may have resulted from camel domestication (Gubser et al, 2004), and whooping cough may have jumped to us from sheep or pigs (Weiss, 2001). Leprosy may have originated in water buffalo (McMichael, 2001) and human influenza may have only started about 4500 years ago with the domestication of waterfowl (Shortridge, 2003). Rhinovirus, the cause of the human cold, may have come from cattle (Rodrigo and Dopazo, 1995). Indeed, before domestication, the common cold may have been common only to them.

Over the last few decades, there has been a dramatic resurgence in emerging infectious diseases, approximately three quarters of which are thought to have come from the animal kingdom (Woolhouse and Gowtage-Sequeria, 2005). The World Health Organization defined the term 'zoonoses' to describe this phenomenon (Mantovani, 2001), from the Greek *zoion* for 'animal' and *nosos* for 'disease'. This trend of increasing zoonotic disease emergence is expected to continue (WHO/FAO/OIE, 2004), and the US Institute of Medicine suggests that without appropriate policies and actions, the future could bring a 'catastrophic storm of microbial threats' (Smolinski et al, 2003).

Animals were domesticated thousands of years ago, though. What new changes are taking place at the human/animal interface that may be responsible for this resurgence of zoonotic disease in recent decades?

In 2004, a joint consultation was convened by the World Health Organization, the Food and Agriculture Organization of the United Nations and the World Organization for Animal Health to elucidate the major drivers of zoonotic disease emergence (WHO/FAO/OIE, 2004). A common theme of primary risk factors for both the emergence and spread of zoonoses was 'increasing demand for animal protein', associated with the expansion and intensification of animal agriculture.

Strep. suis

In 2005, China, the world's largest producer of pork (RaboBank International, 2003), suffered an unprecedented outbreak in scope and lethality of *Streptococcus suis*, a newly emerging zoonotic pig pathogen (Gosline, 2005). *Strep. suis* is a common cause of meningitis in intensively farmed pigs worldwide and presents most often as meningitis in people as well (Huang et al, 2005), particularly those who butcher infected pigs or later handle infected pork products (Gosline, 2005). Due to the involvement of the auditory nerves connecting the inner ears to the brain, half of the human survivors become deaf (Altman, 2005).

The WHO reported that it had never seen such a virulent strain (Nolan, 2005) and blamed intensive confinement conditions as a predisposing factor in its sudden emergence, given the stress-induced suppression of farmed pigs' immune systems (WHO, 2005). The US Department of Agriculture explains that these bacteria can exist as a harmless component of a pig's normal bacterial flora, but stress due to factors such as crowding and poor ventilation can drop the animal's defences long enough for the bacteria to become invasive and cause disease (USDA and Animal and Plant Health Inspection Service, 2005). China's Assistant Minister of Commerce admitted that the disease was 'found to have direct links with the foul environment for raising pigs' (China View, 2005). The disease can spread through respiratory droplets or directly via contact with contaminated blood on improperly sterilized castration scalpels, tooth-cutting pliers or tail-docking knives (Du, 2005). China boasts an estimated 14,000 confined animal feeding operations (CAFOs) (Nierenberg, 2005), colloquially known as factory farms, which tend to have stocking densities conducive to the emergence and spread of disease (Arends et al, 1984). Recent reports of Ebola virus infection in pigs may underscore this point (Barrette et al, 2009).

Nipah virus

This *Strep. suis* outbreak followed years after the emergence of the Nipah virus on an intensive industrial pig farm in Malaysia. Nipah turned out to be one of the deadliest of human pathogens, killing 40 per cent of those infected, a toll that propelled it on to the US list of potential bioterrorism agents (Fritsch, 2003). This virus is also noted for its 'intriguing ability' to cause relapsing brain infections in some survivors (Wong et al, 2002) many months after initial exposure (Wong et al, 2001). Even more concerning, a 2004 resurgence of Nipah virus in Bangladesh showed a case fatality rate on par with Ebola – 75 per cent – and showed evidence of human-to-human transmission (Harcourt et al, 2004). The Nipah virus, like all contagious respiratory diseases, is a density-dependent pathogen (US CIA, 2006). 'Without these large, intensively managed pig farms in Malaysia', the director of the Consortium for Conservation Medicine said, 'it would have been extremely difficult for the virus to emerge' (Nierenberg, 2005).

Bovine spongiform encephalopathy

Global public health experts have identified specific 'dubious practices used in modern animal husbandry' beyond the inherent overstocking, stress and unhygienic conditions that have directly or indirectly launched deadly new diseases (Phua and Lee, 2005). One such 'misguided' practice was the feeding of slaughterhouse waste, blood and excrement to farm animals to save on feed costs (Stapp, 2004).

A leading theory on the origin of BSE, also known as mad cow disease, is that cattle (naturally herbivores) became infected by eating diseased sheep (Kimberlin, 1992). Corporate agribusiness fed protein concentrates (or 'meat and bone meal', euphemistic descriptions of 'trimmings that originate on the killing floor, inedible parts and organs, cleaned entrails, foetuses...') (Ensminger, 1990) to dairy cows to increase milk production (Flaherty, 1993), as well as to most other farm animals (*Economist*, 1990). According to the WHO, nearly 10 million metric tons of slaughterhouse waste was fed to farm animals every year (WHO/OIE, 1999). The recycling of the remains of infected cattle into cattle feed was likely what led to the British mad cow epidemic's explosive spread (Collee, 1993) to nearly two dozen countries around the world in the subsequent 20 years (USDA and Animal and Plant Health Inspection Service, 2005). Dairy producers could have used only corn or soybeans as a protein feed supplement, but slaughter plant by-products can be cheaper (Albert, 2000).

Multidrug-resistant bacteria

Another risky industrial practice is the mass feeding of antibiotics to farm animals. The Union of Concerned Scientists estimate that up to 70 per cent of antimicrobials used in the US are utilized as feed additives for chickens, pigs and cattle for non-therapeutic purposes (Mellon et al, 2001). Indeed, the use of growth-promoting antibiotics in industrial animal agriculture may be responsible for the majority of the increases in antibiotic-resistant human bacterial illness (Tollefson et al, 1999), the emergence of which is increasingly being recognized as a public health problem of global significance (Moore et al, 2006). The worldwide spread of a multidrug-resistant strain of *Salmonella* has been blamed on the use of antibiotics in global fish farming (Angulo and Griffin, 2000).

Alarmingly high rates of methicillin-resistant *Staphylococcus aureus* (MRSA) detection in farm animals and retail meat in Europe, for example, have led to increased scrutiny of the agricultural use of antibiotics. The then Dutch Agriculture, Nature and Food Standards Minister, Cees Veerman, was reported as saying that 'the high usage of antibiotics in livestock farming is the most important factor in the development of antibiotic resistance, a consequence of which is the spread of resistant microorganisms (MRSA included) in animal populations' (Soil Association, 2007). The recent discovery of MRSA in a significant proportion of pigs tested in North America suggests the potential public health risk attributed to farm animal-associated MRSA may be a global phenomenon (Goldburg et al, 2008; Smith et al, 2009a). Long-distance live animal transport and antibiotic use in pigs have both been considered factors in the emergence and spread of livestock associated MRSA (Wulf and Voss, 2008).

Bird flu

The dozens of emerging zoonotic disease threats must be put into context. SARS, which emerged from the live animal meat markets of Asia (Lee and Krilov, 2005), infected thousands of humans and killed hundreds. Nipah infected hundreds and killed scores. *Strep. suis* infected scores and killed dozens. AIDS, which arose from the slaughter and consumption of chimpanzees (Hahn et al, 2000), has infected millions, but only one virus is known to be able to infect billions – influenza.

Influenza, once called the 'last great plague of man' (Kaplan and Webster, 1977), is the only known pathogen capable of a truly global catastrophe (Silverstein, 1981). Unlike other devastating infections like malaria, which is confined equatorially, or HIV, which is only fluid-borne, influenza is considered by Keiji Fukuda, of the US Centers for Disease Control and Prevention, to be the only pathogen carrying the potential to 'infect a huge percentage of the world's population inside the space of a year' (Davies, 1999). In its 4500 years of infecting humans since the first domestication of wild waterbirds, influenza has always been one of the most contagious pathogens (Taylor, 2005). Only since 1997, with the emergence of the highly pathogenic strain H5N1, has it also emerged as one of the deadliest.

H5N1 has so far only killed a few hundred people (WHO, 2009). In a world in which millions die of diseases like malaria, tuberculosis and AIDS, why is there so much concern about bird flu?

The risk of a widespread influenza pandemic is dire and real because it has happened before. An influenza pandemic in 1918 became the deadliest plague in human history, killing up to 100 million people around the world (Johnson and Mueller, 2002) and the 1918 flu virus was probably a bird flu virus (Belshe, 2005) that made more than one quarter of the world's population ill and killed more people in 25 weeks than AIDS has killed in 25 years (Barry, 2004). Despite the harrowing effects of that influenza nearly a century ago, the case mortality rate was less than 5 per cent (Frist, 2005). H5N1, in comparison, has so far officially killed half of its human victims (WHO, 2009).

Free-ranging flocks and wild birds have been blamed for the recent emergence of H5N1, but people have kept chickens in their backyards for thousands of years, and birds have been migrating for millions. What has changed in recent years that led us to this current crisis? According to Dr Robert Webster, the 'godfather of flu research', it is because

> *Farming practices have changed. Previously, we had backyard poultry...* *Now we put millions of chickens into a chicken factory next door to a pig factory, and this virus has the opportunity to get into one of these chicken factories and make billions and billions of these mutations continuously.* *And so what we've changed is the way we raise animals... That's what's changed* (Council on Foreign Relations, 2005).

The United Nations specifically calls on governments to fight what they call 'factory farming':

> *Governments, local authorities, and international agencies need to take a greatly increased role in combating the role of factory farming [which, combined with live bird markets] provide ideal conditions for the virus to spread and mutate into a more dangerous form* (UN, 2005).

Factory farms can be thought of as incubators for the original emergence of dangerous strains of the influenza virus.

Swine flu

The H1N1 swine flu virus has infected millions of people and killed thousands (Centers for Disease Control and Prevention, 2009). Pregnant women and young people are among the hardest hit. Where did this virus come from, and what can be done to help prevent the emergence of flu pandemics in the future?

The genetic fingerprint of the H1N1 swine flu virus confirms that the main ancestor of the 2009 pandemic virus is the triple hybrid human/pig/ bird flu virus that emerged and spread throughout factory farms in the United States more than a decade ago (Smith et al, 2009b). Swine flu emerged in 1918 and remained stable for 80 years in North America until August 1998, when our first hybrid swine flu virus was detected on a factory farm in Sampson County, North Carolina (Wuethrich, 2003), the county with the single highest pig population in the country, confining more than 2 million pigs (Sampson County Health Department, 2007).

The factory farm in which the virus was first found was a breeding facility confining thousands of sows in sow stalls, metal cages so small the pigs can't even turn around. These instruments of extreme confinement have not only been criticized as inhumane, but may pose a public health threat, as crated sows have been reported to have higher stress levels and impaired immune systems (Siegel, 1983).

Other factors that make intensive farms such breeding grounds for disease include the sheer numbers of animals (Poljak et al, 2008), the over-crowding (Maes et al, 2000), the millions of gallons of excrement, which releases ammonia that burns the animals' lungs (Donham, 1991) and the lack of adequate fresh air and sunlight. The ultraviolet (UV) rays in sunlight are actually quite effective in destroying these viruses. Just 30 minutes in direct sunlight completely inactivates the flu virus, but it can last for days in the shade and weeks in moist manure (Songserm et al, 2006). Over-crowding thousands of animals snout to snout into stressful filthy football-field sized sheds may create a Perfect Storm environment for the emergence and spread of new 'superstrains' of influenza (Greger, 2007a).

The public health community has been warning about the human health risks posed by factory farms for years. In 2003, the American Public Health Association, the largest affiliation of public health professionals in the world, called for a moratorium on factory farms (American Public Health Association, 2003). In 2008, the Pew Commission on Industrial Farm Animal Production released its final report and concluded that industrialized animal agriculture posed 'unacceptable' public health risks (Pew Commission on Industrial Farm Animal Production, 2008a). The Pew Commission was a prestigious, independent panel chaired by a former Kansas Governor and including a former US Secretary of Agriculture, former Assistant Surgeon General, and the Dean of the University of Iowa College of Public Health. The Commission recommended that gestation crates be banned, noting that '[p]ractices that restrict natural motion, such as sow gestation crates, induce high levels of stress in the animals and threaten their health, which in turn may threaten our health' (Pew Commission on Industrial Farm Animal Production, 2008b).

The worst case scenario might be if the H1N1 swine flu virus combined with the highly pathogenic H5N1 bird flu virus, both of which have been found infecting pigs. If a single pig in parts of Asia or Africa where bird flu has become endemic becomes co-infected with both the new swine flu and bird flu, the concern is that it could theoretically produce a virus with the human transmissibility of swine flu, but also with the human lethality of bird flu (NN, 2009).

Michael Osterholm, the director of the US Center for Infectious Disease Research and Policy and an associate director within the US Department of Homeland Security, tried to describe what an H5N1-type pandemic could look like in one of the leading US public policy journals, *Foreign Affairs*. Osterholm suggested policy makers consider the devastation of the 2004 tsunami in South Asia: 'Duplicate it in every major urban centre and rural community around the planet simultaneously, add in the paralyzing fear and panic of contagion, and we begin to get some sense of the potential of pandemic influenza' (Kennedy, 2005).

'An influenza pandemic of even moderate impact', Osterholm continued, 'will result in the biggest single human disaster ever – far greater than AIDS, 9/11, all wars in the 20th century and the recent tsunami combined. It has the potential to redirect world history as the Black Death redirected European history in the 14th century' (Kennedy, 2005).

Hopefully, the world will move away from raising animals by the billions under intensive confinement, thus potentially lowering the risk of us ever being in this same precarious situation in the future.

Conclusion

H1N1 swine flu is not the first virus whose emergence has been linked to factory farms, and won't be the last unless significant changes are made in the way animals are raised, measures as simple as providing straw bedding so these animals don't have the immune-suppressive stress of lying on bare concrete their whole lives (André and Tuyttens, 2005), have been shown to significantly decrease swine flu transmission rates (Ewald et al, 1994). Such a simple measure, yet we deny these animals even this modicum of mercy – both to their detriment and potentially to ours as well.

In addition to phasing out extreme confinement, the replacement of long distance live animal transport with a carcass-only trade (Greger, 2007b), intensive farms with smaller-scale production with lower densities of animals (Maillard and Gonzalez, 2006) and an overall reduction in the number of animals intensively confined and raised for food (Benatar, 2007) have all been suggested as ways to reduce the emergence of zoonotic infectious diseases.

'The bottom line', according to a spokesperson for the WHO, 'is that humans have to think about how they treat their animals, how they farm them, and how they market them – basically the whole relationship between the animal kingdom and the human kingdom is coming under stress' (Torrey and Yolken, 2005). Along with human culpability, though, comes hope. If changes in human behaviour can cause new plagues, changes in human behaviour may prevent them in the future.

References

Albert, D. (2000) 'EU meat meal industry wants handout to survive ban', *Reuters World Report*, 5 December

Altman, L. K. (2005) 'Pig disease in China worries UN', *New York Times*, available at www.iht.com/bin/print_ipub.php?file=/articles/2005/08/05/news/pig.php, last accessed 5 August 2009

American Public Health Association (2003) 'Precautionary moratorium on new concentrated animal feed operations', available at www.apha.org/advocacy/policy/policysearch/default.htm?id=1243, last accessed 18 November 2009

André, F. and Tuyttens, M. (2005) 'The importance of straw for pig and cattle welfare: A review', *Applied Animal Behaviour Science*, vol 92, pp261–282

Angulo, F. J. and Griffin, P. M. (2000) 'Changes in antimicrobial resistance in *Salmonella enterica* serovar Typhimurium', *Emerging Infectious Diseases*, vol 6, no 4, pp436–437

Arends, J. P., Hartwig, N., Rudolphy, M. and Zanen, H. C. (1984) 'Carrier rate of *Streptococcus suis* capsular type 2 in palatine tonsils of slaughtered pigs', *Journal of Clinical Microbiology*, vol 20, no 5, pp945–947

Armelagos, G. J., Barnes, K. C. and Lin, J. (1996) 'Disease in human evolution: The re-emergence of infectious disease in the third epidemiological transition', *National Museum of Natural History Bulletin for Teachers*, vol 18, no 3

Barrette, R. W., Metwally, S. A., Rowland, J. M., Xu, L., Zaki, S. R., Nichol, S. T., Rollin, P. E., Towner, J. S., Shieh, W., Batten, B., Sealy, T. K., Carrillo, C., Moran, K. E., Bracht, A. J., Mayr, G. A., Sirios-Cruz, M., Catbagan, D. P., Lautner, E. A., Ksiazek, T. G., White, W. R. and McIntosh, M. T. (2009) 'Discovery of swine as a host for the Reston ebolavirus', *Science*, vol 325, pp204–206

Barry, J. M. (2004) 'Viruses of mass destruction', *Fortune*, 1 November

Belshe, R. B. (2005) 'The origins of pandemic influenza – lessons from the 1918 virus', *New England Journal of Medicine*, vol 353, no 21, pp2209–2211

Benatar, D. (2007) 'The chickens come home to roost', *American Journal of Public Health*, vol 97, pp1545–1546

Centers for Disease Control and Prevention (2009) 'CDC estimates of 2009 H1N1 influenza cases, hospitalizations and deaths in the United States, April – October 17, 2009', available at www.cdc.gov/h1n1flu/estimates_2009_h1n1.htm, last accessed 12 November 2009

China View (2005) 'China drafts, revises laws to safeguard animal welfare', available at http://news.xinhuanet.com/english/2005-11/04/content_3729580.htm, last accessed 4 November 2009

Collee, G. (1993) 'BSE stocktaking 1993', *Lancet*, vol 342, no 8874, pp790–793, available at www.cyber-dyne.com/~tom/essay_collee.html, last accessed 9 December 2009

Council on Foreign Relations (2005) 'Session 1: Avian flu – where do we stand?', *Conference on the Global Threat of Pandemic Influenza*, available at http://cfr.org/publication/9230/council_on_foreign_relations_conference_on_the_global_threat_of_pandemic_influenza_session_1.html, last accessed 16 November 2009

Davies, P. (1999) 'The plague in waiting', *Guardian*, available at http://guardian.co.uk/birdflu/story/0,,1131473,00.html, last accessed 7 August 2009

Donham, K. J. (1991) 'Association of environmental air contaminants with disease and productivity in swine', *American Journal of Veterinary Research*, vol 52, pp1723–1730

Du, W. (2005) '*Streptococcus suis*, (*S. suis*) pork production and safety', Ontario Ministry of Agriculture, Food and Rural Affairs

Economist (1990) 'Mad, bad and dangerous to eat?', February, pp89–90

Ensminger, M. E. (1990) *Feeds and Nutrition*, Ensminger Publishing Co., Clovis, CA

Ewald, C., Heer, A. and Havenith, U. (1994) 'Factors associated with the occurrence of influenza A virus infections in fattening swine', *Berliner und Münchener tierärztliche Wochenschrift*, vol 107, pp256–262

Flaherty, M. (1993) 'Mad cow disease dispute. UW conference poses frightening questions', *Wisconsin State Journal*, 26 September, p1C

Frist, B. (2005) 'Manhattan project for the 21st century', Harvard Medical School Health Care Policy Seidman Lecture, available at http://frist.senate.gov/_files/060105manhattan.pdf, last accessed 1 June 2009

Fritsch, P. (2003) 'Containing the outbreak: Scientists search for human hand behind outbreak of jungle virus', *Wall Street Journal*, 19 June

Goldburg, R., Roach, S., Wallinga, D. and Mellon, M. (2008) 'The risks of pigging out on antibiotics', *Science*, vol 321, no 5894, p1294

Gosline, A. (2005) 'Mysterious disease outbreak in China baffles WHO', available at www.newscientist.com/article.ns?id=dn7740, last accessed 9 December 2009

Greger, M. (2007a) 'The human/animal interface: Emergence and resurgence of zoonotic infectious diseases', *Critical Reviews in Microbiology*, vol 33, pp243–299

Greger, M. (2007b) 'The long haul: The risks of livestock transport.' *Biosecurity and Bioterrorism*, vol 5, pp301–311

Gubser, C., Hué, S., Kellam, P. and Smith, G. L. (2004) 'Poxvirus genomes: A phylogenetic analysis', *Journal of General Virology*, vol 85, pp105–117

Hahn, B. H., Shaw, G. M., De Cock, K. M. and Sharp, P. M. (2000) 'AIDS as a zoonosis: Scientific and public health implications', *Science*, vol 287, pp607–614

Harcourt, B. H., Lowe, L., Tamin, A., Liu, X., Bankamp, B., Bowden, N., Rollin, P. E., Comer, J. A., Ksiazek, T. G., Hossain, M. J., Gurley, E. S., Breiman, R. F., Bellini, W. J., Rota, P. A. (2004) 'Genetic characterization of Nipah virus, Bangladesh, 2004. Centers for Disease Control and Prevention', *Emerging Infectious Diseases*, vol 11, no 10, available at www.cdc.gov/ncidod/EID/vol11no10/05-0513.htm, last accessed 9 December 2009

Huang, Y. T., Teng, L. J., Ho, S. W. and Hsueh, P. R. (2005) '*Streptococcus suis* infection', *Journal of Microbiology, Immunology and Infection*, vol 38, pp306–313, available at http://jmii.org/content/abstracts/v38n5p306.php, last accessed 9 December 2009

Johnson, N. P. A. S. and Mueller, J. (2002) 'Updating the accounts: Global mortality of the 1918–1920 "Spanish" influenza pandemic', *Bulletin of the History of Medicine*, vol 76, pp105–115

Kaplan, M. M. and Webster, R. G. (1977) 'The epidemiology of influenza', *Scientific American*, vol 237, pp88–106

Kennedy, M. (2005) 'Bird flu could kill millions: Global pandemic warning from WHO. "We're not crying wolf. There is a wolf. We just don't know when it's coming"', *Gazette* (Montreal), 9 March, pA1

Kimberlin, R. H. (1992) 'Human spongiform encephalopathies and BSE', *Medical Laboratory Sciences*, vol 49, pp216–217

Lee, P. J. and Krilov, L. R. (2005) 'When animal viruses attack: SARS and avian influenza', *Pediatric Annals*, vol 34, no. 1, pp43–52

Maes, D., Deluyker, H., Verdonck, M., Castryck, F., Miry, C., Vrijens, B. and de Kruif, A. (2000) 'Herd factors associated with the seroprevalences of four major respiratory pathogens in slaughter pigs from farrow-to-finish pig herds', *Veterinary Research*, vol 31, pp313–327

Maillard, J. C. and Gonzalez, J. P. (2006) Biodiversity and emerging diseases', *Annals of the New York Academy of Sciences*, vol 1081, pp1–16

Mantovani, A. (2001) 'Notes on the development of the concept of zoonoses', WHO Mediterranean Zoonoses Control Centre Information Circular 51, available at www.mzcp-zoonoses.gr/pdfen/circ_51.pdf, last accessed 9 December 2009

McMichael, T. (2001) *Human Frontiers, Environments and Disease*, Cambridge University Press, Cambridge

Mellon, M. G., Benbrook, C. and Lutz Benbrook, K. (2001) *Hogging It! Estimates of Antimicrobial Abuse in Livestock*, Union of Concerned Scientists, Cambridge, MA

Moore, J. E., Barton, M. D., Blair, I. A., Corcoran, D., Dooley, J. S. G., Fanning, S., Kempf, I., Lastovica, A. J., Lowery, C. J., Matsuda, M., McDowell, D. A., McMahon, A., Millar, B. C., Rao, J. R., Rooney, P. J., Seal, B. S., Snelling, W. J. and Tolba, O. (2006) 'The epidemiology of antibiotic resistance in Campylobacter', *Microbes and Infection*, vol 8, pp1955–1966

Nierenberg, D. (2005) 'Happier meals: Rethinking the global meat industry', Worldwatch Paper 171, available at www.worldwatch.org/pubs/paper/171/, last accessed 9 December 2009

NN (2009) 'Exclusive: SARS sleuth tracks swine flu, attacks WHO,' *ScienceInsider*, 4 May, available at http://blogs.sciencemag.org/scienceinsider/2009/05/exclusive-meet. html, last accessed 9 December 2009

Nolan, T. (2005) '40 people die from pig-borne bacteria', available at www.abc.net.au/am/content/2005/s1441324.htm, last accessed 9 December 2009

Pew Commission on Industrial Farm Animal Production (2008a) 'Expert panel highlights serious public health threats from industrial animal agriculture', available at www.pewtrusts.org/news_room_detail.aspx?id=37968, last accessed 11 April 2009

Pew Commission on Industrial Farm Animal Production (2008b) 'Putting meat on the table: Industrial farm animal production in America', Executive summary, p13, available at www.ncifap.org/_images/PCIFAPSmry.pdf, last accessed 9 December 2009

Phua, K. and Lee, L. K. (2005) 'Meeting the challenges of epidemic infectious disease outbreaks: An agenda for research', *Journal of Public Health Policy*, vol 26, pp122–132

Poljak, Z., Dewey, C. E., Martin, S. W., Christensen, J., Carman, S. and Friendship, R. M. (2008) 'Prevalence of and risk factors for influenza in southern Ontario swine herds in 2001 and 2003', *Canadian Journal of Veterinary Research*, vol 72, pp7–17

RaboBank International (2003) 'China's meat industry overview: Quietly moving towards industrialization', *Food and Agribusiness Research*, May

Rodrigo, M. J., and Dopazo, J. (1995) 'Evolutionary analysis of the Picornavirus family', *Journal of Molecular Evolution*, vol 40, pp362–371

Sampson County Health Department (2007) 'Community health assessment: Sampson County', available at www.sampsonnc.com/Health Assessment.pdf, last accessed 9 December 2009

Shortridge, K. F. (2003) 'Severe acute respiratory syndrome and influenza', *American Journal of Critical Care and Respiratory Medicine*, vol 168, pp1416–1420

Siegel, H. S. (1983) 'Effects of intensive production methods on livestock health', *Agro-Ecosystems*, vol 8, pp215–230

Silverstein, A. M. (1981) *Pure Politics and Impure Science, the Swine Flu Affair*, Johns Hopkins University Press, Baltimore, MA, pp129–131

Smith, T. C., Male, M. J., Harper, A. L., Kroeger, J. S., Tinkler, G. P., Moritz, E. D., Capuano, A. W., Herwaldt, L. A. and Diekema, D. J. (2009a) 'Methicillin-resistant *Staphylococcus aureus* (MRSA) strain ST398 is present in midwestern US swine and swine workers', *PLoS One*, vol 4, e4258

Smith, G. J., Vijaykrishna, D., Bahl, J., Lycett, S. J., Worobey, M., Pybus, O. G., Ma, S. K., Cheung, C. L., Raghwani, J., Bhatt, S., Peiris, J. S. M., Guan, Y. and Rambaut, A. (2009b) 'Origins and evolutionary genomics of the 2009 swine-origin H1N1 influenza A epidemic', *Nature*, vol 459, pp1122–1125

Smolinksi, M. S., Hamburg, M. A. and Lederberg, J. (2003) *Microbial Threats to Health: Emergence, Detection and Response*, National Academies Press, Washington, DC

Soil Association (2007) 'MRSA in farm animals and meat', available at www.soilassociation.org/LinkClick.aspx?fileticket=%2bmWBoFr348s%3d&tabid=385, last accessed 9 December 2009

Songserm, T., Jam-On, R., Sae-Heng, N. and Meemak, N. (2006) 'Survival and stability of HPAI H5N1 in different environments and susceptibility to disinfectants', in Schudel, A. and Lombard, M. (eds) *Proceedings of the OIE/FAO International Scientific Conference on Avian Influenza*, vol 124, Karger, Switzerland, International Association for Biologicals

Stapp, K. (2004) 'Scientists warn of fast-spreading global viruses', *IPS-Inter Press Service*, 23 February

Taylor, M. (2005) 'Is there a plague on the way?', *Farm Journal*, 10 March

Tollefson, L., Fedorka-Cray, P. J. and Angulo, F. J. (1999) 'Public health aspects of antibiotic resistance monitoring in the USA', *ACTA Veterinaria Scandinavica Supplement*, vol 92, pp67–75

Torrey, E. F. and Yolken, R. H. (2005) *Beasts of the Earth: Animals, Humans, and Disease*, Rutgers University Press, New Jersey

UN (2005) 'UN task forces battle misconceptions of avian flu, mount Indonesian campaign', UN News Centre, available at www.un.org/apps/news/story.asp?NewsID =16342&Cr=bird&Cr1=flu, last accessed 24 October 2009

US CIA (2006) *CIA World Fact Book*, Malaysia, available at http://cia.gov/cia/publications/factbook/geos/my.html, last accessed 29 March 2009

USDA, Veterinary Services, Center for Emerging Issues (2005) '*Streptococcus suis* outbreak, swine and human, China: Emerging disease notice', available at www. aphis.usda.gov/vs/ceah/cei/taf/emergingdiseasenotice_files/strep_suis_china.htm, last accessed 9 December 2009

USDA and Animal and Plant Health Inspection Service (2005) 'List of USDA-recognized animal health status of countries/areas regarding specific livestock or poultry diseases', USDA and Animal and Plant Health Inspection Service, USA

Weiss, R. A. (2001) 'Animal origins of human infectious disease, The Leeuwenhoek Lecture', *Philosophical Transactions of the Royal Society of London, Series B, Biological Sciences*, vol 356, pp957–977

Wong, K. T., Shieh, W. J., Zaki, S. R. and Tan, C. T. (2002) 'Nipah virus infection, an emerging paramyxoviral zoonosis', *Springer Seminars in Immunopathology*, vol 24, pp215–228

Wong, S. C., Ooi, M. H., Wong, M. N. L., Tio, P. H., Solomon, T. and Cardosa, M. J. (2001) 'Late presentation of Nipah virus encephalitis and kinetics of the humoral immune response', *Journal of Neurology, Neurosurgery & Psychiatry*, vol 71, pp552–554

Woolhouse, M. E. and Gowtage-Sequeria S. (2005) 'Host range and emerging and reemerging pathogens', *Emerging Infectious Diseases*, vol 11, pp1842–1847

WHO/OIE (1999) 'WHO Consultation on Public Health and Animal Transmissible Spongiform Encephalopathies: Epidemiology, Risk and Research Requirements', 1–31 December, Geneva, Switzerland

WHO/FAO/OIE (2004) Report of the WHO/FAO/OIE joint consultation on emerging zoonotic diseases, available at www.whqlibdoc.who.int/hq/2004/WHO_CDS_CPE_ ZFK_2004.9.pdf, last accessed 10 December 2009

WHO (2005) 'Streptococcus suis fact sheet', available at www.wpro.who.int/media_ centre/fact_sheets/fs_20050802.htm, last accessed 10 December 2009

WHO (2009) 'Cumulative number of confirmed human cases of avian influenza A(H5N1)', www.who.int/csr/disease/avian_influenza/country/en/, last accessed 8 April 2009

Wuethrich, B. (2003) 'Infectious disease: Chasing the fickle swine flu,' *Science*, vol 299, pp1502–1505

Wulf, M. and Voss, A. (2008) 'MRSA in livestock animals – an epidemic waiting to happen?', *Clinical Microbiology and Infection*, vol 14, pp519–521

11

Environmentally Sustainable and Equitable Meat Consumption in a Climate Change World

Anthony J. McMichael and Ainslie J. Butler

Introduction

As we humans settle in to our traverse through the 21st century, the consumption of meat is rising worldwide, both because of the continuing growth in human numbers and because of the emerging demand for meat and animal foods that accompanies urbanization, economic development and middle-class consumerism. However, tensions persist: access to animal foods remains very unequal, much commercial livestock production is environmentally damaging – including a growing contribution to global greenhouse gas emissions – and moral dilemmas abound in relation to the treatment of animals.

In this chapter, we explore the prospects for achieving an environmentally sustainable and socially equitable regime of meat production and consumption. The interplay between environmental, economic, social, cultural and moral issues is complex, and much of the research and policy challenge does not reduce to simple bottom-line causal attribution and effect estimation. Further, conclusions and related policies will necessarily vary between populations and locations, and will change over time as technologies and circumstances evolve. Our conclusions are therefore mostly indicative rather than definitive, and we offer them with caution.

Meat has been part of the human diet since the *Homo* genus emerged a little over 2 million years ago. Over the ensuing millennia the capture of animals and scavenging and consumption of meat, and, much later, the intake of dairy foods, influenced many aspects of the evolution of both human biology and culture. Meat provides energy, protein and a range of micronutrients at much higher concentrations than do plant foods. In addition to the attractions of taste, texture and concentrated energy content, red meat is a valuable source of the unsaturated omega-3 fatty acids (the misnamed 'fish oils'), iron, zinc, amino acids and vitamin B12.

Palaeo-anthropological studies have documented this long pre-history of meat eating (McMichael, 2001). From australopithecine gatherers to gatherer-scavengers, our ancient forebears then became hominine gatherer-hunters and, in some thinly vegetated regions (e.g. high latitude Arctic), predominantly hunter-gatherers. The evidence includes indicative evolutionary changes in human dentition and jaws from late australopithecine times, systematic cut-marks on fossilized animal bones dating from early *Homo* times and variations in isotope ratios in human fossil bones reflecting trends in relative amounts of animal and plant foods in the diet. The much later advent of nomadic animal herding ensured a reliable source of meat in the diet. Then, appearing first in southwest Asia (the Levant region) from around 8000 years ago, settled agrarian living that included domesticated livestock led to the consumption of milk and other dairy products (Ponting, 2007) – and thus providing a more sustained and energy-efficient source of animal foods. In the modern world, animal-source foods are eaten widely, across continents, cultures and economic strata – although the ratio of animal to plant foods varies greatly.

As societies modernize, the accompanying 'nutrition transition' entails an increase in consumption of meat, especially red meat (Popkin, 2003). There is, however, a serious environmental 'downside' to much of the production of meat and other animal-source foods, especially the high-volume intensified commercial form of that production. The substantial local, regional and, now, global environmental damage that is being caused by the world's livestock sector is of growing concern (Steinfeld et al, 2006). Today, an especially urgent environmental concern is the large contribution (18 per cent) that the overall livestock sector makes to the global emission of greenhouse gases (GHGs) and hence to human-driven climate change (Steinfeld et al, 2006; McMichael et al, 2007). Currently that sector, including land use practices to produce animal feed, accounts for 20–25 per cent of global greenhouse gas emissions, expressed as carbon dioxide equivalents, or CO_2e (Stern, 2006; Solomon et al, 2007). Meanwhile, an indicative (though off-beat) statistic is that the production of one kg of beef can generate as much carbon dioxide (CO_2) as driving 250 kilometres in the average European car or using a 100-watt light bulb continuously for around 20 days (Ogino et al, 2007).

Livestock husbandry methods in many countries have become increasingly intensive over the past half-century, as, indeed, has much cereal crop production in the hands (increasingly) of the transnational corporate sector. Water demand by the livestock sector is huge, surface soils are often physically damaged, land clearing for grazing and feed production is widespread, much manure and other waste is generated and air pollution is a frequent community respiratory health hazard in populations surrounding concentrated animal feeding operations – which is well documented for intensive hog farming in North Carolina, USA (Wing and Wolf, 2000). Around two thirds of all agricultural land (approximately 2 billion of the 3 billion available hectares) and one third of land surface is used, directly or indirectly, in livestock production (Steinfeld et al, 2006). Much of this use is still of a traditional and non-intensive kind, with relatively low environmental impact. However, to feed expanding urban populations with rising dietary expectations of animal-source foods, further growth and, most probably, intensification of production methods appears inevitable over coming decades. Can this be achieved sustainably?

Constraining the environmental and climatic impacts of livestock production will need to draw on three main strategies: (i) a reduction in the total global consumption of animal-source foods, (ii) changes in livestock husbandry methods and (iii) changes in the overall profile of preferred animal-food species and strains. The first of those strategies can be pursued both by changes in dietary culture and preference and, of course, by the accelerated attainment of replacement-level fertility in order to limit further world population growth. The latter is happening with increasing rapidity, at last; however the former will present further challenge until a 'tipping point' in public awareness and engagement is reached.

The environmental damage being caused by today's livestock sector is part of the wider problem of the increasingly commercialized and industrialized global food system that is now overloading many of nature's absorptive and regenerative capacities. The International Assessment of Agricultural Knowledge, Science and Technology for Development, coordinated by the World Bank, concluded that much of the world's current food production, particularly 'industrial' mono-cultural production practices such as the conversion of large swathes of the Brazilian Amazon to broad-acres soybean production, is not sustainable (World Bank and UN World Food Program, 2008). Related criticisms had previously been made of the environmentally damaging and depleting consequences of the 'Green Revolution' that transformed cereal grain yields in much of Asia in the 1970s and 1980s (Ponting, 2007).

This precarious modern situation is the culmination of a long historical struggle to produce sufficient food to feed family groups, communities, city-states and whole populations. The following brief historical review should help to clarify the contemporary choices we have in relation to food production methods, food security and environmental sustainability. These choices entail other criteria, especially for those for whom meat consumption poses philosophical, moral and religious dilemmas – important issues, but not the subject of this chapter (but see Chapters 13 and 14 in this book). Even so, a serious ethical issue remains because of our moral responsibilities to environmental and ecological systems, and, by extension to future generations and the greater goal of environmental sustainability (Ilea, 2009).

World food situation: Environmental resource base and food insecurity

Today, food availability in developed countries is widely taken for granted. Food shortages and famines belong (at least for the moment) in the history books. Yet an abundance and diversity of food is a remarkable recent circumstance. Indeed, the year-round constancy of food supply and nutrient content is somewhat at variance with our ancestral seasonally dominated hunter-gatherer past and hence with aspects of human biological evolution.

Food shortages (both seasonal and episodic), nutrient deficiencies and periodic famines have long beset human groups. Many of our behavioural tendencies and metabolic characteristics – including the craving for sugar and fat, overeating in response to visual cues and the conservation of insulin resistance as trade-off for preserved fertility under conditions of food shortage – are the evolutionary legacy of survival-enhancing selection pressures from that more precarious ancestral era (McMichael, 2001; Corbett et al, 2009). Those tendencies, in modern, wealthy, food-secure societies, underlie much of the non-communicable chronic disease burden that now prevails in middle and later adulthood.

There is, however, no continuing guarantee of this abundance of food. Indeed, the year 2008 generated considerable alarm at the emergence of widespread food shortage. Food prices (especially of rice and other cereal grains) rose dramatically as farm production levels, global stocks and the international trade in essential grains declined. That episode may have been an early indicator that, for the first time at the *global* level, humans in aggregate are now reaching environmental limits to the ready achievement of further increases in food production (Butler, 2009; Hall and Day, 2009).

The global human population has increased from 1 billion in the 1820s to 6.8 billion by 2010. A total population of 8–9 billion is projected by 2050. Life expectancies have increased in most regions, as have levels of wealth and consumption. However, this has been accompanied by steadily increasing evidence that humanity is now living beyond Earth's carrying capacity. We are, in effect, subsidizing much economic growth by depleting the planet's natural environmental stocks – soils, fisheries, fresh water supplies, forests, biodiversity – and by overloading its capacity to absorb wastes. This, too, underlies the emergence of climate change, depletion of stratospheric ozone, ocean acidification and accelerating loss of biodiversity. In addition to widespread mismanagement of fresh water supplies, the melting of alpine glaciers due to global warming will further reduce river flows in many great river basins and food-producing regions. Meanwhile, as economic globalization proceeds, the replacement of small farm-holdings and of crop diversity with industrial monocultural farming has diminished the genetic versatility and the resilience of world agriculture as it faces these great modern environmental and climatic stresses (Wright, 2004; World Bank and UN World Food Program, 2008).

The world map of livestock production and consumption, and of human-driven environmental changes, shows great differences. International data show a positive correlation between national per capita income and daily average food energy intake (McMichael et al, 2007). High-energy diets typically comprise elevated levels of animal-source foods and lower levels of grains and complex carbohydrates (including in fruits and vegetables). In nutrient terms, as income rises so too does the intake of energy-dense sugars, total fat and animal fat. Figure 11.2, adapted from the 2006 FAO Report, shows recent and projected trends in the consumption of animal-source foods by major region of the world. In absolute terms, especially in

Box 11.1 *Populations, environment and food: A perennial struggle that now faces global limits*

The advent of meat eating has influenced aspects of human biological evolution. In the 1920s, Raymond Dart, in Johannesburg, showed from fossil evidence (the famous Taung Child) that the transition from australopithecine to hominine, around 2.5 million years ago, entailed changes in the lower jaw and dentition, presumably reflecting dietary shifts as more meat was scavenged and less grinding of fibrous plan foods was required (McMichael, 2001). The 'expensive tissue hypothesis' (Aiello and Wheeler, 1995) posits that the subsequent increase in energy-dense meat consumption, via active hunting, lessened the need for a large and hard-working intestine for the digestion and fermentation of a large volume of plant foods. Since the gut is a metabolically expensive organ, this dietary shift in the early *Homo* genus would have released precious metabolic energy that could, via random Darwinian 'experiments', be reapplied with benefit to another very energy-intensive organ, the brain. On this hypothesis, as the human gut contracted, the brain was able to expand.

That larger brain subsequently potentiated improved hunting methods, and the exploitation of diverse environments. A major step forward occurred with the control of fire, from up to a million years ago. This enabled meat and fibrous tubers to be cooked, chewed and digested, which increased the environmental carrying capacity and therefore population size. Further, an increase in amylase enzyme levels appears to have occurred, in Darwinian response to cooked starchy foods (Perry et al, 2007). This would have further reduced the need for colonic fermentation.

The typical paleolithic hunter-gatherer diet had a high content of meat and protein (Figure 11.1).

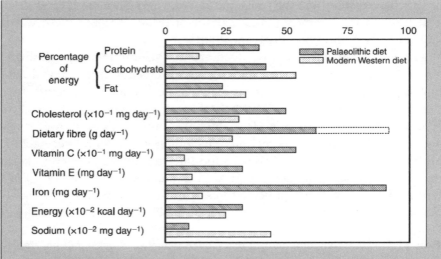

Source: McMichael (2001)

Figure 11.1 Comparison of the palaeo-skeletal (and other) estimates of the typical hunter-gatherer diet with that of modern western populations

From around 11,000 years ago, early farming emerged, livestock were later domesticated, food supplies increased (in return for harder work), populations expanded and food surplus allowed social stratification and hierarchical privilege. Food trade between early city-states allowed yet further regional population increases. Farm yields, however, were precarious, vulnerable to weather reversals and to eventual over-exploitation of forest, soil and water (McMichael, 2001).

Those trends in food production and associated environmental pressures have accelerated markedly in recent times. Human numbers have increased eight-fold in the past two centuries. Thus, the wheel has come full circle: the ancient dietary shift that boosted our brain capacity has led to a modern world in which there are too many of us wanting to eat too much food, including, increasingly, an excess of meat. Global livestock production is projected to double by 2050, outstripping the rest of the agricultural sector. This trend is not sustainable (Steinfeld et al, 2006).

relation to projected future increases, the East Asian region (with a major contribution from China) is becoming increasingly prominent.

The annual production of meat will double from 229 million tonnes in 1999–2001 to 465 million tonnes in 2050, according to FAO estimates; and milk output will increase from 580 million tonnes to 1043 million tonnes

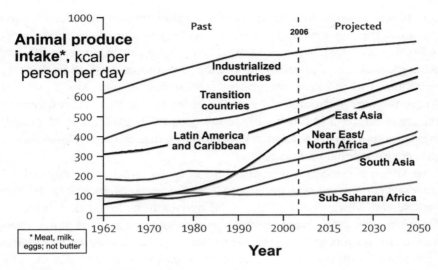

Source: Adapted from FAO (2006)

Figure 11.2 Time-trends in per-person consumption of animal foods, past and projected, for major categories and regions

(Steinfeld et al, 2006). As Figure 11.2 shows, the projected increase is predominantly in low- or middle-income countries.

Environmental impacts of livestock production

The herding and breeding of animals for human consumption, particularly in intensified high-volume commercial settings, puts many pressures on the natural environment, including problems of over-grazing damage to pastures and pollution of waterways. Such environmental impacts were unusual before the 20th century – and remain limited in many traditional livestock-raising cultures where animals often feed on plant sources, and on terrain, unavailable to humans.

Modern 'industrial' meat production requires large inputs of energy, water and cereal grains. For example, the water demands are reflected in the fact that American and European diets, typically high in meat consumption, require inputs of approximately 5000 litres of water daily – compared with the largely vegetarian-based diets of Africa and Asia that require only 2000 litres of water per day. As meat consumption levels escalate in many developing regions, agricultural water requirements will increase and pressures on water systems will intensify. Such methods typically cause much damage to local environments: pollution via effluent and chemical (especially fertilizer) run-off, soil erosion and water depletion. Further, there is considerable inefficiency in the conversion of animal feed energy to animals-as-food energy (Tilman et al, 2002). While the figures vary significantly between different farming cultures and methods, up to nine units of non-pasture feed energy are required to produce one unit of feedlot beef energy. The equivalent figures for pigs and poultry are of the order of 4:1 and 2:1, respectively (see also Chapter 3 in this book). To the extent that the feeding of animals diverts foods away from human consumption, that efficiency loss detracts from environmental sustainability.

As stated previously, up to one third of global greenhouse gas (GHG) emissions come from agriculture and land use activities, including the livestock sector. More than half that total derives from livestock production (Steinfeld et al, 2006; Stern, 2006) – an amount similar to the global emissions from the transport sector. Greenhouse gas emissions from the livestock sector comprise around one tenth of global emissions of CO_2, plus 35–40 per cent of methane emissions and 65 per cent of nitrous oxide. Both methane (CH_4) and nitrous oxide (N_2O) have a much greater global warming potential, by unit volume, than does CO_2.

The livestock sector's major contributor to emissions is the enteric methane from ruminant (digastric) grazers: cattle, sheep and goats, as well as buffaloes and camels. These digastric animals have evolved a digestive system that can break down the cellulose from grasses in their fore-stomachs, via bacterial action, with the production of methane that is then regurgitated. A typical cow regurgitates around 700 litres of methane per day. Compared with the warming effect of CO_2 over 20 years, by volume, methane is more than 70 times more powerful in its heat-trapping and warming action.

It is important to acknowledge, here, that these estimations of the local and global environmental impacts of livestock production are complex (Garnett, 2009). For example, the fact that grazing ruminants can often convert vegetation growing on otherwise non-arable land into food, as meat, for humans represents an increase in the (human) carrying capacity of that particular environment. Further, animals contribute to our wealth and health in ways other than the production of meat (which, in environmental and energy-use terms, is usually the most wasteful of all uses of animals). Livestock animals, variously in different settings, eat wastes (food scraps, cellulosic products, etc.), contribute labour and may help to maintain the vitality of some ecosystems.

A second example comes from the environmental arithmetic of grain-fed livestock. The production and transport of cereal grains for livestock is energy-intensive. China, once an exporter of soybean products, now imports almost half of Brazil's soybean harvest as feed for cattle and pigs. Further, the antecedent land clearing is a source of CO_2 emissions. Yet grain-fed ruminants produce less methane, per individual animal, than do grazing ruminants (Steinfeld et al, 2006) – since there is much less cellulose in grains to be broken down in the ruminant fore-stomach than there is in grasses.

Intensive livestock production systems are also associated with a higher risk of transmission through the food chain of infectious disease from stressed and infected animals, in addition to the increased probability of emergence of novel zoonotic infectious agents – discussed further below (see also Chapter 10 in this book). This risk of infectious food-poisoning and the preceding examples highlight the complexity of conducting a full 'life cycle analysis' of the environmental credits and debits of livestock production. Clearly, simply shifting from one mode of agricultural practice to another will not simply and automatically rectify the issues of environmental impacts and pressures.

Solutions to many of these excessive and growing environmental impacts of the livestock sector will require substantial changes in economic policy, culture and consumer behaviour – such as shifts in population dietary patterns that increase the proportion of animal protein from poultry, monogastrics and plant-eating fish. Meanwhile, some of the environmental pressures can be alleviated via available technologies and strategies. For example, GHG emissions could be reduced by up to 20 per cent per livestock unit, at relatively low cost, by combinations of the following options (McMichael et al, 2007):

- reduction of atmospheric CO_2 through reforestation, reduced deforestation and restoration of organic carbon to soils and pastures, plus – if feasible – carbon sequestration (in other words, 'CO_2 capture and storage');
- reduced release of methane from enteric fermentation in livestock through improved feed efficiency;
- improved management practices for manure and biogas to mitigate methane release;
- more efficient utilization of nitrogenous fertilizers to mitigate release of nitrous oxide.

Livestock production as source of infectious disease in humans

For around 6000–7000 years, traditional small farms accommodated mixed species living closely with humans – goats, pigs, cattle, ducks, geese, chickens and perhaps a water buffalo or donkey (Ponting, 2007). Accordingly, there has long been animal-to-human transfer of (often novel) infectious agents. Indeed, this zoonotic route accounts for the origins of many familiar textbook diseases such as measles, mumps, tuberculosis, cholera and chicken pox (McNeill, 1976).

Today's intensified animal production facilitates the emergence of new infectious diseases – which, of course, is a continuation of the ancient narrative whereby most of the textbook infectious diseases in today's human populations originated from early agrarian exposures to domesticated animals (Weiss and McMichael, 2004). The modern extension of this narrative has been well illustrated by the 'mad cow' episode in the UK of the 1980s with its 1990s human equivalent 'variant Creutzfeldt-Jakob Disease'; the perennial generation of new influenza strains in association with poultry production in East and Southeast Asia (including traditional pig-duck-human contacts in villages in southern China and high-density chicken farming in Southeast Asia), and the emergence of the Nipah virus via intensified pig farming in Malaysia. The Nipah virus, originating from displaced food-deprived forest bats, was able to infect battery-farmed pigs and thereby find its way into human animal handlers and workers, killing approximately half of the more than 200 infected persons (Chua et al, 2000).

Cross-infection also occurs when animal species are sold together in the marketplace. The 1997 outbreak of avian influenza in Hong Kong occurred in mixed markets where live chickens, quail and ducks were stacked together in close quarters with humans (Webster, 2004). The taste for exotic animal species exacerbates the risk of exposure to new microbes. This pathway probably triggered the severe acute respiratory syndrome (SARS) epidemic, in which (wild) palm civet cats, traded across national borders, were the prime suspect as the source of the new corona virus that caused SARS (Weiss and McMichael, 2004).

The infectious diseases risks to human health also include the generation of antibiotic-resistant bacteria within the livestock sector. Around 70 per cent of all antibiotics used in the US today are fed to livestock to prevent infections and to enhance growth – a practice that is now banned in some countries. This practice has been implicated in the emergence and spread of some types of antibiotic-resistant bacteria, including the worrisome vancomycin-resistant enterococcus.

Meat consumption and the risk of non-infectious diseases

Nutrition scientists typically recommend an individual intake of around 50–100g of meat per day, to enhance the diet, ensure adequate protein intake and provide sufficient iron and vitamin B12. Average meat intake in the high-income world is well above that level, within the range of 200–300g per day. The US has the highest per-person daily intake. In contrast, the average intake in sub-Saharan Africa is around one-eighth of the American level, typically within the range of 20–40g per day (McMichael et al, 2007). Some of this difference reflects cultural preference; much of it reflects access and affordability – an issue of inequity.

The case for radically modifying livestock production to meet the needs of environmental sustainability, especially given the sector's significant contribution to climate change, is eliciting new and intensive scrutiny. Such analysis should take into account the public health risks attendant on diets high in meat and animal food content.

Epidemiological evidence shows that an excessive consumption of red meat in the high-income western diet leads to a heightened risk of premature mortality, including from cardiovascular disease. A recent ten-year follow-up study of half a million Americans aged 50–71 found an increased risk in overall mortality in persons consuming high levels of red meat compared with those consuming at low levels – a 31 per cent increase in risk in males and a 36 per cent increase in females. The increased risk in high-consumers was evident in deaths from both cardiovascular diseases and cancer (Sinha et al, 2009).

The risks of obesity and heart (coronary artery) diseases derive most probably from the high content of saturated fat in domesticated livestock meat. The high saturated fat content derives both from selective breeding over many millennia (although undergoing some reversal in recent decades) and from the relative physical inactivity of those animals – especially where grain-fed.

There is persuasive epidemiological evidence that the risk of large bowel cancer is raised at significantly elevated levels of red meat consumption – and perhaps particularly for meat grilled or braised via direct heat at high temperatures. The World Cancer Research Fund, in its summary of a comprehensive and intensive review of world scientific literature on diet and cancer, estimated a 30 per cent increase in risk of colorectal cancer for every 100 gm of red meat consumed per day (WCRF/AICR, 2007). Consumption of processed meats also appears to increase the risk of colorectal cancer, potentially via a different carcinogenic mechanism. Whether the risks of breast and several other cancers are also increased by red meat consumption remains unresolved.

Meanwhile, a modest increase in the level of consumption of animal products in low-intake populations should provide benefits to health, particularly in the antenatal period (mother and fetus) and in early childhood (growth and development – physical and intellectual). Low levels of animal protein intake, in the context of poor availability of a range of nutritionally balanced foodstuffs, also contributes to iron deficiency and malnutrition, which could be addressed by moderate increases in consumption of red meat.

Box 11.2 *Health benefits of climate-friendly reduced meat consumption in the UK and Brazil*

An international collaborative research programme was conducted in 2009 to investigate the health co-benefits of various climate change mitigation actions. As part of that research, the health benefits of a reduction in animal product consumption to reduce greenhouse gas emissions in the UK and Brazil were estimated, focusing mainly on the risk of coronary heart disease (Friel et al, 2009).

The estimated benefit of reducing the intake of saturated fat and cholesterol at the level that would result from reduced animal-foods consumption in accord with national GHG emissions reduction targets, could save 16 per cent of years of life lost from ischaemic heart disease in the UK, and 17 per cent of years of life lost from ischaemic heart disease in Brazil. These modelled estimates assume a 30 per cent reduction in production and consumption of livestock by 2030, along with technological practice changes, in order to reduce GHG emissions to meet the agreed emissions reductions targets.

This evidence of the various health consequences of meat consumption points to the population health benefits that would result from reducing red meat consumption in those (mostly wealthy) populations that currently have high levels of consumption. Further, a reduction in meat consumption in these countries would allow modest increases in consumption levels in developing countries, without increasing the total global contribution of livestock to greenhouse gas emissions. This would allow low-consuming developing countries to increase their meat and calorie consumption towards a generally healthier level.

Contraction and convergence: A sustainable and equitable resolution

The composite issue of achieving modest, equitable and environmentally sustainable animal-source foods consumption can be best tackled via the principle of 'contraction and convergence'. This intuitively appealing strategy originated with contraction-and-convergence modelling of future changes in per capita GHG emission levels, to ensure contraction of total global emissions via convergence on a common level of per capita emission (Meyer, 2000). This strategy has gained increasing recognition, stimulating its suggested application to solving the historically based disparities between high- and low-income countries for many greenhouse-related production and consumption practices.

In the present context, then, the 'contraction' component requires that the world commit to reducing the global total intake of meat (especially red meat from ruminant animals) by some specified date in future. This would, ideally, be part of a portfolio strategy to mitigate climate change – across various sectors such as commerce, energy generation, urban planning and human behaviour. The 'convergence' component specifies that the contraction of consumption is achieved equitably. Thus, today's high-consuming populations would reduce their intake downwards to the agreed global per capita level of consumption, while low-consuming populations would increase their intake up to, but not beyond, that agreed level.

The first detailed study of this strategy, taking account of the projected rise in world population and its regional composition, proposed an appropriate global average target figure of 90g of meat per day – with not more than 50g as red meat from ruminant animals (McMichael et al, 2007). There are, of course, many non-ruminant sources of meat. The long-established dietary traditions in much of East and Southeast Asia have relied much on chickens as a source of protein, and, in non-Muslim cultures, on pigs (but see Chapters 8 and 9 in this book for information on the health and welfare problems associated with modern chicken and pork production). Many populations have potential access to other, often healthier, sources of meat. In Australia, for example, both the environment and the nation's health would benefit from a change to kangaroo meat – lean meat, essentially devoid of saturated fat, but containing health-promoting omega-3 fatty acids (albeit at lesser concentrations than in fish), but the ethical issues of slaughter methods could make this a problematic issue.

The gains to the environment from such reductions in meat consumption would accrue both immediately and in terms of longer-term sustainability (including stabilization of the global climate). Replacing ruminant red meat with meat from monogastric animals (poultry, marsupials, other mammals) or from vegetarian fish-farming would, respectively, reduce methane emissions from ruminants and carbon dioxide release from energy-inefficient processes on land, and reduce pressures on wild fisheries as sources of fish-meal for aquaculture. Reduction of meat consumption would also help to free up, for human consumption, some of the crops used as animal feed. (A similar argument, of alienation of human food sources, is now being made in relation to some of the feedstock produced for biofuels.)

Overall, then, this contraction and convergence strategy, in relation to world meat consumption, would be a win–win strategy. Greenhouse gas emissions would be reduced and global warming would be slowed. Health risks in high-consuming populations would be lowered, while gains in nutritional status would occur in lower-income countries via reduction of deficiencies of iron, protein and energy intake – with benefits to child health and development in particular. In lower-income countries, the 'convergence' ceiling figure of 90g of meat per day should mostly preclude any increases in risks of cancer, heart disease or obesity-related diabetes. This contraction-and-convergence strategy, phased in over several decades, would therefore be good for the planet, good for enhancing global equity, and generally good for population health.

Box 11.3 *Examples of government initiatives towards reducing dietary GHG emissions*

Governments can promote policies that promote both environmentally sustainable practices and increases in human health. Recent examples include the following:

- During the January 2009 'Grüne Woche' (Green Week) event in Berlin, the German federal environmental agency called for a reduction in mean consumption of ruminant red meat to prewar levels in response to concerns over the environmental impact of the livestock industry. Germany currently ranks among the highest in Europe for red meat consumption

- In the UK, the National Health Service (NHS) announced in January 2009 its decision to promote meat-free menus at hospitals in an attempt to cut GHG emissions.

- In May 2009, the Belgian city of Ghent declared that it would be promoting meat-free days every Thursday, under the banner 'Donderdag: Veggie Dag' to combat obesity, global warming and animal cruelty. They also intend to extend 'Veggie Dag' to include schoolchildren, providing vegetarian food as the default meal in the cafeterias. Other towns in Europe and in Canada have expressed growing interest in promoting the concept.

- In New Zealand, Estonia and the US, legislation has been proposed to apply carbon (or methane) taxes to the livestock industry. In 2003 in New Zealand the rationale behind this proposed 'flatulence tax' was to fund research on reducing livestock emissions through technology in light of complying with the Kyoto Protocol. This legislation was effectively blocked by agricultural groups. However, New Zealand is now making efforts to phase out intensive pig farming. The story in Estonia is similar, where the livestock tax proposed by the government was shelved due to lobbying from the agricultural industry in 2008.

These examples illustrate a range of policy avenues for addressing the issue of an environmentally sustainable diet and the equity-promoting principle of global contraction and convergence. Promoting actions at personal, city-wide or national levels can help to promote awareness of the issue at global level – and, eventually, effective response to it at that level.

Conclusion

The world's livestock sector is at the junction of several of the great environmental and moral issues of the modern age. This includes the urgent issues of food insecurity, under-nutrition and its health consequences, environmental degradation, exacerbation of global climate change and concern for animal welfare. Recognition of these issues, and the consequent enlightened social responses, could lead to: (i) greater equity of food availability and nutritional outcomes between regions and populations, (ii) a more compassionate and ecologically attuned livestock sector and (iii) restoration of damaged environments and, crucially, a slowing of climate change.

Meat consumption is rising rapidly in many countries, along with industrialization, urbanization and the emerging might of middle-class consumer preference. Currently, there are great differences between countries in levels of meat consumption per person. Some of this difference reflects cultural preference; some reflects unequal access and affordability. The latter represents a moral challenge. A further moral issue, and political challenge, bearing on future generations arises from the fact that, with current production methods, the projected level of meat consumption over coming decades is not environmentally sustainable. In particular, the global livestock sector is a major contributor to climate change.

Climate change, with its many recognized and growing risks to human health and survival, is today's most clearly defined and recognized global environmental problem. Meanwhile, the world's struggle to achieve and maintain universal food security in face of the still-expanding global population underscores the great systemic challenge that we face globally, in seeking a sustainable way to live and to manage this planet's natural environment.

The world community must strive for a sustainable and more equitable 'world diet' by combining a stabilization and then reduction of globally averaged meat production per person, while also achieving a much greater equality of access to affordable meat and other animal-source foods between and within populations. This goal and strategy will benefit both human health and the environment, and will help redirect the human species onto a sustainable path – as we continue our traverse of this most challenging of centuries.

References

Aiello, L. C. and Wheeler, P. (1995) 'The expensive-tissue hypothesis: The brain and the digestive system in human and primate evolution', *Current Anthropology*, vol 36, p199

Butler, C.D. (2009) 'Food security in the Asia-Pacific: Malthus, limits and environmental challenges', *Asia Pacific Journal of Clinical Nutrition*, vol 18, no 4, pp577–584

Chua, K. B., Bellini, W. J., Rota, P. A., Harcourt, B. H., Tamin, A., Lam, S. K., Ksiazek, T. G., Rollin, P. E., Zaki, S. R., Shieh, W. J., Goldsmith, C. S., Gubler, D. J., Roehrig, J. T., Eaton, B., Gould, A. R., Olson, J., Field, H., Daniels, P., Ling, A. E., Peters, C. J., Anderson, L. J. and Mahy, B. W. J. (2000) 'Nipah virus: A recently emergent deadly paramyxovirus', *Science*, vol 288, pp1432–1435

Corbett, S. J., McMichael, A. J. and Prentice, A. M. (2009) 'Type 2 diabetes, cardiovascular disease and the evolutionary paradox of the polycystic ovary syndrome: A fertility first hypothesis', *American Journal of Human Biology*, vol 21, pp587–591, doi: 10.1002/ajhb.20937

Friel, S., Dangour, A., Garnett, T., Lock, K., Chalabi, Z., Roberts, I., Butler, A., Butler, C., Waage, J., McMichael A. J. and Haines, A. (2009) 'Public health benefits of strategies to reduce greenhouse-gas emissions: Food and agriculture', *Lancet*, vol 374, no 9706, pp145–152

Garnett, T. (2009) 'Livestock-related greenhouse gas emissions: Impacts and options for policy makers', *Environmental Science and Policy*, vol 12, pp491–503

Hall, C. A. S. and Day, J. W. (2009) 'Revisiting the limits to growth after peak oil in the 1970s a rising world population and the finite resources available to support it were hot topics. Interest faded – but it's time to take another look', *American Scientist*, vol 97, pp230–237

Ilea, R. C. (2009) 'Intensive livestock farming: Global trends, increased environmental concerns, and ethical solutions', *Journal of Agricultural and Environmental Ethics*, vol 22, pp153–167

McMichael, A. J. (2001) *Human Frontier, Environments and Disease: Past Patterns, Future Uncertainties*, Cambridge University Press, Cambridge

McMichael, A. J. (2005) 'Integrating nutrition with ecology: Balancing the health of humans and biosphere', *Public Health Nutrition*, vol 8, pp706–715

McMichael, A. J., Powles, J. W., Butler, C. D. and Uauy, R. (2007) 'Food, livestock production, energy, climate change, and health', *Lancet*, vol 370, pp1253–1263

McNeill, W. H. (1976) *Plagues and Peoples*, Anchor Press, Doubleday, New York

Meyer, A. (2000) *Contraction and Convergence. The Global Solution to Climate Change: Schumacher Briefings 5*, Greenbooks for the Schumacher Society, Bristol, USA

Ogino, A., Orito, H., Shimada, K. and Hirooka, H. (2007) 'Evaluating environmental impacts of the Japanese beef cow-calf system by the life cycle assessment method', *Animal Science Journal*, vol 78, pp424–432

Perry, G. H., Dominy, N. J., Claw, K. G., Lee, A. S., Fiegler, H., Redon, R., Werner, J., Villanea, F. A., Mountain, J. L., Misra, R., Carter, N. P., Lee, C. and Stone, A. C. (2007) 'Diet and the evolution of human amylase gene copy number variation', *Nature Genetics*, vol 39, pp1256–1260

Ponting, C. (2007) *A New Green History of the World. The Environment and the Collapse of Great Civilisations*, Vintage Books, London

Popkin, B. M. (2003) 'The nutrition transition in the developing world', *Development Policy Review 2003*, vol 21, pp581–597

Sinha, R., Cross, A. J., Graubard, B. I., Leitzmann, M. F. and Schatzkin, A. (2009) 'Meat intake and mortality: A prospective study of over half a million people', *Archives of Internal Medicine*, vol 169, pp562–571

Solomon, S., Qin, D., Manning, M., Marquis, M., Averyt, K., Tignor, M. M. B., Miller, H. L. R. and Chen, Z. (eds) *Climate Change 2007: The Physical Science Basis.*

Contribution of Working Group I to the Fourth Assessment Report of the Intergovernmental Panel on Climate Change, Cambridge University Press, Cambridge

Steinfeld, H., Gerber, P., Wassenaar, T., Castel, V., Rosales, M. and de Haan, C. (2006) 'Livestock's long shadow: Environmental issues and options', FAO, Rome

Stern, N. (2006) *The Economics of Climate Change: The Stern Review*, Cambridge University Press, Cambridge

Tilman, D., Cassman K. G., Pamela A., Matson P. A., Naylor, R. and Polasky, S. (2002) 'Agricultural sustainability and intensive production practices', *Nature*, vol 418, pp671–677

WCRF/AICR (2007) 'Food, nutrition, physical activity and the prevention of cancer: A global perspective', World Cancer Research Fund/American Institute of Cancer Research, Washington DC

Webster, R. G. (2004) 'Wet markets – a continuing source of severe acute respiratory syndrome and influenza?', *Lancet*, vol 363, pp234–236

Weiss, R. and McMichael, A. J. (2004) 'Social and environmental risk factors in the emergence of infectious diseases', *Nature Medicine*, vol 10, ppS70–76

Wing, S. and Wolf, S. (2000) 'Intensive livestock operations, health, and quality of life among eastern North Carolina residents', *Environmental Health Perspectives*, vol 108, pp233–238

World Bank and UN World Food Program (2008) '*International Assessment for Agricultural Knowledge, Science and Technology for Development Report*', World Bank, Washington DC

Wright, R. (2004) *A Short History of Progress*, Anansi, Toronto

12

How Much Meat and Milk is Optimal for Health?

Mike Rayner and Peter Scarborough

Introduction

Meat and dairy products (MDPs) have long been regarded as good for health (Fiddes, 1991) but this assumption is increasingly being called into question. It has come to be recognized that vegetarians are generally as healthy as non-vegetarians and sometimes healthier. This has led some to question whether MDPs are necessary for a healthy diet and even whether diets would be healthier if people ate less MDPs than they currently do. The debate about MDPs and health is particularly intense in industrialized countries, where people have diets that are relatively high in MDPs and where it has come to be realized that the production of MDPs – at least as generally practiced in several countries – has detrimental effects on the environment (see Chapter 11).

This chapter aims to summarize where the debate about the health benefits (and otherwise) of MDPs has reached. First, we look briefly at what we mean by 'meat and dairy products'. Second, we look at why MDPs have until recently generally been thought to be good for health. Third, we look at some of the recent research that questions the view that MDPs are good for health. As part of the examination we look at trends in meat and dairy consumption in the UK and elsewhere and the health effects of these trends. Fourth, we look at what official bodies are precisely saying about MDP consumption. And finally we look at how we might decide the answer to the question that is the title for this chapter: 'How much meat and milk is optimal for health?' but also how we might decide the equally important question: 'What type of MDPs are optimal for health?'

What are meat and dairy products?

By 'MDPs' we mean all edible products derived from mammals or birds. So by 'meat' we mean all types of red meat (e.g. from cows, sheep and pigs), white meat (e.g. from chicken), processed meat and offal, but we do not include fish or shellfish. By 'dairy products' we mean milk, cream and processed dairy products such as butter, cheese and yoghurt. Within 'MDPs' we include eggs. 'MDPs' therefore include a large number of different types of foods with different nutritional properties and hence different effects on health.

The animals that end up being consumed as meat or that are the source of dairy products and eggs are fed and kept in different ways. Some have suggested that the ways animals are farmed have important effects on the healthiness of their products and we look briefly at this issue below. More-over MDPs can be subject to extensive processing and this too can have a major impact on their health-related properties. Again this is something we examine below. This means that as well as the question: 'How much meat and dairy is optimal for health?' we also need to ask: 'What types of MDPs are optimal for health?'

Perceptions of the health benefits of MDPs

The idea that MDPs are basically good for us, including our health, has, until recently, been deeply entrenched within our culture. Meat – for many people – has had to be a part of their main meal of the day. In a study carried out in the mid 1980s 200 women were asked what the family needs to eat properly and the researchers found that 'meat was mentioned by the women more frequently than any other food. In fact only five women thought meat was not an important item of the family diet' (Kerr and Charles, 1986, cited in Fiddes, 1991). Attitudes to the relationship between meat and health may be changing but not very quickly, as evidenced by trends in vegetarianism (if anything declining) and in meat consumption (if anything increasing) (Food Standards Agency and Department of Health, 2010).

Dairy products, particularly milk, have also been considered by most people to be an essential part of a healthy diet perhaps even more so than meat. As a *Times* editorial of 1936 put it: 'Milk makes you sleep o'nights, gives you a milky complexion, makes muscle, gives you a healthy old age, and makes the toddler king of the castle. What more can men and women want?' (quoted in McKee, 1997). Note, too, the outrage that was generated by Margaret Thatcher's decision as Secretary of State for Education, in 1970, to end free school milk for primary school children over the age of seven: a decision for which she was dubbed 'Milk Snatcher' (Webster, 1997). Milk and dairy products are still regarded as health-promoting foods in many quarters.

Obviously there have long been vegetarians. The Vegetarian Society was founded – for a variety of reasons including the purported health benefits of vegetarianism – in 1847 (Vegetarian Society, n.d). But vegetarians have always been in the minority and remain so today. Current estimates of vegetarianism in the UK suggest that 2–5 per cent of the adult population are vegetarians (defined here as individuals who consume dairy products and eggs but no meat or fish) and less than 1 per cent are vegans (defined here as individuals who do not consume any meat, fish, dairy products or eggs) (Food Standards Agency and Department of Health, 2004, 2010).

The popular belief that MDPs are good for health has, until relatively recently, been supported by nutritional orthodoxy. Furthermore much effort has been expended by the nutritional establishment, backed by the livestock industry, on informing the public that meat is an indispensible part of a healthy diet. The primary reason for this is that meat is relatively high in protein, minerals such as iron and zinc and vitamins such as vitamin B12 and milk and dairy products are high in protein and calcium.

Protein – like other nutrients within foods – is essential for the normal growth and development of the human body. But protein is often regarded as the most important of these nutrients. The very word 'protein' comes from the Greek word πρωτειος (*proteios*), meaning 'primary'.

The recognition that protein is essential for normal growth and development has led – not particularly logically – to the belief that there is a need for large quantities of protein in the human diet. This is sometimes known as the 'protein myth'. This myth/belief can be traced back at least as far back as the work of Baron Justus von Liebig – the 19th-century German chemist. In *Animal Chemistry* (1846) and *Researches on the Chemistry of Food* (1847), von Liebig glorified meat at the essential source of material to replenish muscular strength lost during physical activity. He believed (wrongly as it turns out) that the more physically active a person, the more protein they need. The doctrines of von Liebig were expanded upon by his many followers including Lyon Playfair in the UK, Carlo von Voit and Max Ruber in Germany and Wilbur Atwater in the US, all of whom made estimates of human protein requirements that were accepted as the basis of government food policy until well into the 20th century (Cannon, 2003). These estimates were based largely not on what people needed, but on what the average healthy person at that time consumed and so tended to overestimate people's actual requirements for protein. Since then estimates of requirements have become more sophisticated and are now based more closely on need.

So today the UK Government's estimates that the average women requires 36g of protein per day and the average man 44g (Department of Health, 1991). However, average protein intakes are currently around 66g for an adult woman and 88g for an adult man, in other words nearly twice as much as 'required' (Food Standards Agency and Department of Health, 2004). Moreover these high average intake levels mean that virtually no one in the UK suffers from a protein deficiency. Note, however, that older people and women who are pregnant or breastfeeding need more protein than the average person when measured as a percentage of energy intake (WHO, 2007).

Perhaps surprisingly the health effects of eating less protein than deemed essential are poorly understood. This is because low protein diets tend to be generally nutrient-poor, deficient to varying degrees in a range of other nutrients, and also often associated with other environmental factors that can adversely influence health (WHO, 2007).

The health effects of very high – or even moderately high – protein intakes are even less clear. What is clear is that high protein diets promote the growth of young children so that as adults they are much taller than they would otherwise be. It seems likely that the steady increase in average height in most developed countries over the 20th century (in many cases starting well before that) has been mainly due to increasing protein intakes. Furthermore differences in protein intakes are largely responsible for the differences in heights between nationalities. But faster growing and taller is not necessarily healthier and some have argued that high protein diets have led to all sorts of health and other social problems (Cannon, 2003).

While most foods contain some protein, MDPs tend to be relatively high in protein compared with plant-based foods. This means that MDPs are the biggest source of protein in the UK diet: meat providing 38 per cent and dairy products 13 per cent (Food Standards Agency and Department of Health, 2010). This does not mean that plant-based foods could not provide enough protein if people ate less MDPs. In fact the average person is already getting about 31g a day of protein from cereals, fruit, vegetables (including potatoes) and nuts, and if we assume that official recommendations are right and the average person needs about 40g a day, then we don't need to eat much more of these foods to meet the recommendation.

Protein from MDPs is sometimes considered to be of 'higher quality' than protein from plant sources for two reasons. First, the protein in meat and dairy protein is said to provide a 'better balance' of the amino acids that make up proteins. Some amino acids are essential in that they cannot be made within the human body. Meat, and to a less extent dairy products, tend to have all these essential amino acids in fairly high and similar amounts while plant-based foods may not. However plant-based diets (provided that they are varied enough) can provide all the essential amino acids the body needs. Second, the protein in MDPs is thought to be more digestible than protein from plant-based sources, although not much more (Millward and Garnett, 2010).

It's not just protein that is found in high levels in MDPs but also other nutrients – particularly minerals and vitamins – that are essential for growth and for the maintenance of normal function. According to the UK Food

Table 12.1 *Most important of the vitamins and minerals provided by MDPs in the UK diet*

Nutrient	% total intake from meat	% of UK population below LRNI	Nutrient	% total intake from dairy	% of UK population below LRNI
Iron	85	13	Calcium	43	4
Zinc	34	4	Iodine	38	3
Niacin	34	1	Vitamin B12	36	1
Vitamin B12	30	1	Riboflavin	33	5
Vitamin A	28	7	Phosphorus	24	0
Vitamin D	22	No official LRNI	Zinc	17	4

Note: LRNI: the government's 'Lower reference nutrient intake'.

Source: Food Standards Agency and Department of Health (2004)

Standards Agency: 'Meat is a good source of … vitamins and minerals, such as iron, selenium, zinc, and B vitamins. It is also one of the main sources of vitamin B12' and 'Milk and dairy products … are great sources of … vitamins A and B12. They're also an important source of calcium' (Food Standards Agency, n.d. 1). MDPs are also good sources of iodine, riboflavin and niacin (see Table 12.1).

The vitamins and minerals that are present in high levels in MDPs all have different functions. For example iron is an essential component of haemoglobin, which transports oxygen around the body in the blood stream. Calcium is an essential component of bone. Vitamin B12 is needed for normal neural development, and so on.

However, in many instances, the precise relationship between these minerals and vitamins and health is still not well understood. It is clear that extremely low levels of the nutrients lead to disease. For example a very low iron intake leads to anaemia, which produces fatigue and impaired performance in adults, reduced motor skills in children and even death. But the effects of moderately low levels are less clear.

In the UK the government has published what it considers to be adequate levels for mineral and vitamin intakes in the population (Department of Health, 1991) These adequate levels (Lower Reference Nutrient Intakes) are set relative to what is considered desirable (Estimated Average Requirements). Table 12.1 shows that there are very few people in the UK who have inadequate levels of the minerals and vitamins that are present in high levels in MDPs, the exception being iron where there is a problem with girls and young women.

But as with protein it is unclear whether the average intakes of the vitamins and minerals found in high levels in MDPs are higher or lower than necessary for optimal health. It is of course clear that there are some people in the UK who have less than desirable or even adequate levels of the minerals and vitamins found in high levels in MDPs (as published by the government). But the precise health effects of this are difficult to quantify and may not be that large.

Recent research questioning the health benefits of MDPs

The mid 20th-century consensus that MDPs were good for health because they were high in protein and essential minerals and vitamins began to break down in the 1960s and 1970s with the recognition that the high and increasing rates of many chronic diseases in developed countries were partly diet related.

Cardiovascular disease

One of the first relatively convincing studies in this area was carried out in the late 1970s and early 1980s by the pioneering American epidemiologist Ancel Keys, who, by comparing rates of cardiovascular disease (CVD) (that is, coronary heart disease (CHD) and stroke) in seven different countries, showed that there was a strong association between average CVD rates and average blood cholesterol levels. Average blood cholesterol levels were in turn found to be positively associated with average levels of dietary cholesterol and of saturated fat and possibly negatively associated with average levels of unsaturated fat (though this was less clear).

The results of these cross-country comparisons were quickly confirmed by observational studies on individuals and more latterly in experimental studies (including randomized controlled trials) and thereby shown to be causal relationships and not just associations (Skeaff and Miller, 2009).

Whereas almost all of the fat in plant-based foods is unsaturated (either monounsaturated or polyunsaturated) a substantial amount of the fat in MDPs is saturated. At the very least 60 per cent of saturated fat in the UK diet comes from MDPs (Food Standards Agency and Department of Health, 2010). Furthermore it is only animal-derived products that contain dietary cholesterol. Therefore, consumption of MDPs – particularly MDPs high in saturated fat – has come to be linked with an increased risk of CVD.

Research carried out since the early 1980s on the relationship between MDP consumption and risk of CVD has shown it to be more complicated than previously thought. First, it turned out that saturated fat was more important than dietary cholesterol in raising risk of blood cholesterol and then CVD. Then different studies generated confusing results about the importance of high intakes of total fat (as opposed to whether the fat is saturated or unsaturated) and the relative importance of low intakes of monounsaturated and polyunsaturated fat (as opposed to high intakes of saturated fat). It was also shown that different types of saturated fat (and indeed polyunsaturated fats) were more important than others in their impact on risk of CVD.

The polyunsaturated fat story has become particularly complicated. Some polyunsaturated fats (linoleic acid and α-linolenic acid) cannot be formed in the body and thus it is 'essential' that the diet supplies enough of them. The UK Government therefore recommends a minimum intake of 1 per cent of energy from linoleic and similar n-6 (polyunsaturated) fats and 0.2 per cent of energy from α-linolenic and similar long chain n-3 (polyunsaturated) fats (Department of Health, 1991). But since the 1960s and 1970s the evidence has grown that higher levels of both n-3 polyunsaturated fats (found in large amounts in fish but also in vegetable oils such as rape seed oil) and n-6 polyunsaturated fats (found in large amounts in vegetable oils such as sunflower and corn oils) reduce the risk of CVD in different ways and particularly when replacing saturated fats (Skeaff and Miller, 2009).

Since the 1980s evidence has also grown that trans-fats formed from unsaturated fats ultimately derived from plant sources either in animal rumens (which then find their way into MDPs) or industrially by the process known as hydrogenation, are particularly toxic in relation to risk of CVD (Mozaffarian et al, 2006). Ruminant trans-fats and industrially produced trans-fats have similar metabolic effects but it has generally been thought that the latter can be more easily removed from the food supply than the former.

Despite all this recent research the basic story has remained essentially the same for around the last 20 or 30 years. High intakes of saturated fat increase the risk of CVD and a reduction in average saturated fat intake in the UK and elsewhere would generate substantial health benefits. We calculate that reducing current average intakes from approx 14 per cent of energy (as at present) to 10 per cent of energy (the UK Government's recommended goal) would save over 4000 lives a year in the UK (Scarborough et al, in preparation). Given that MDPs are the main source of saturated fat and that the health benefits of the current high intakes of MDPs uncertain there would seem – on this basis alone – to be an obvious case for reducing meat and dairy intake. However, most nutritionists have been reluctant to draw a direct connection, arguing that it is only those MDPs that are high in saturated fat that constitute the risk, not high MDP consumption as such.

There has also been a tendency for all the research into the relationship between different types of fats and CVD to obscure the basic story. There is currently some debate about whether the fat composition of MDPs can be enhanced in relation to its effects on health by different feeding practices. For example it is certainly the case that n-3 polyunsaturated fat levels in beef can be increased by high forage-based diets (Givens, 2005). This is likely to mean that the meat from grass-fed cattle will be slightly healthier than from non-grass-fed cattle. However, it seems unlikely that the slightly improved fat composition of beef from grass-fed animals (compared for example with beef produced more intensively) will have any major impact on the health of its consumers.

Research on the relationships between different types of dietary fat, blood cholesterol levels and risk of CVD has been accompanied by research on the relationship between diet, blood pressure levels and risk of CVD. Even before it became clear that a raised blood cholesterol level increased the risk of CVD it was known that raised blood pressure did so and it was suspected that raised blood pressure was partly diet-related. The clearest dietary cause of high blood pressure has turned out to be high salt intakes (Strazzullo et al, 2009).

Only a small amount of salt (more precisely sodium) is required for normal body function and this can be obtained easily enough through its natural occurrence in many foods including meat, fish, eggs, fruits and vegetables. However, salt is used as a preservative and flavour enhancer in most processed foods and consumption of these processed foods can lead to excessive consumption of sodium. Much of the MDP consumption in the UK comes in the form of processed foods (cheese, bacon, ham, sausages, ready meals, etc.) and 40 per cent of the salt in the UK diet comes from such MDPs (Food Standards Agency and Department of Health, 2010). We estimate that reducing average salt intakes in the UK from current levels at around 9g per day to the recommended 6g per day would save about 8000 lives a year (Scarborough et al, in preparation).

An important paper has just been published that systematically reviews all the good quality studies of the relationship between red and processed meat consumption and the risk of CVD (Micha et al, 2010). It shows clearly that risk of CVD (and diabetes) is greater for processed meat than for unprocessed (red) meat consumption, but it should not be taken to suggest that there is no increased risk from consuming unprocessed meat.

Cancer

As with relationships between diet and CVD, the relationship between diet and cancer has become increasingly clear over the last few decades – due particularly to the work of the World Cancer Research Fund (WCRF) who have systematically reviewed all the available studies and published two seminal reports (WCRF/AICR, 1997, 2007).

In the process the WCRF have examined all possible relationships between MDP consumption and cancer. The one relationship that the WCRF describes as 'convincing' is the relationship between red and processed meat intake and colorectal cancer. A number of potential mechanisms have been suggested, including the formation of carcinogens in the high temperature cooking of meat, and the potential damage to cells in the gut of free radicals associated with the haem iron component of red meat. At present, these mechanisms are not well established.

Colorectal cancer kills about 13,000 people each year in the UK (Allender et al, 2008) and only a small proportion of these could be prevented by reducing red and processed meat consumption in the UK. The number of deaths from colorectal cancer prevented would be far short of the 12,000 deaths from cardiovascular disease that could be avoided by reducing saturated fat and salt levels to government recommended levels.

Obesity

It is well known that obesity is increasing and some have suggested that this is somehow associated with the high consumption of MDPs. However, there is no substantial evidence for any relationship. Some MDPs are energy dense and the WCRF suggests that the evidence linking the consumption of large amounts of energy dense foods with overweight and obesity is 'probable'. But the WCRF also indicates that the evidence linking MDP consumption with overweight and obesity is 'limited and inconclusive' (WCRF/AICR, 2007).

Vegetarianism and veganism

Given the negative consequences of MDP consumption and the relatively few and uncertain benefits of a high consumption as described above, it is perhaps not surprising that vegetarians and vegans are often healthier than the general population (Key et al, 1999; Mann, 2009). And it is also clear from this that MDPs are not essential elements of a healthy diet for all individuals. However, a note of caution is needed here. Vegetarians and vegans tend to be more health conscious than the general population, which may explain some of the differences in health outcomes between the two groups. For example as well as eating fewer MDPs, they eat more fruit and vegetables, they smoke less and do more physical activity. Although many people have no problem living healthily on vegan and vegetarian diets, others have found these diets more difficult to maintain.

Trends in MDP consumption

It is an assumption of this chapter that diet is an important determinant of health. Estimates vary but we find that around 30,000 deaths a year could be avoided (or at least delayed) if people in the UK ate more healthily (Scarborough et al, in preparation). MDPs provide a major source of calories to the average diet in many countries. On this basis alone MDP consumption is likely to have important effects on various aspects of health (both positive and negative).

This is borne out by recent studies of health trends in different countries including the UK. Figures 12.1 and 12.2 show trends in meat and dairy consumption over the last 40 years. Over this period there has been little change in overall death rates from cancer but there has been a significant fall in deaths from CVD. For example CVD death rates in adults aged under 75 have fallen by over 60 per cent since 1970 (Allender et al, 2008). This has been partly due to changes in diet and in particular the reduced consumption of high fat dairy products – principally whole milk and butter. People have switched from whole milk to semi-skimmed milk (Figure 12.2) and from butter to polyunsaturated spreads. Both of these 'switches' have

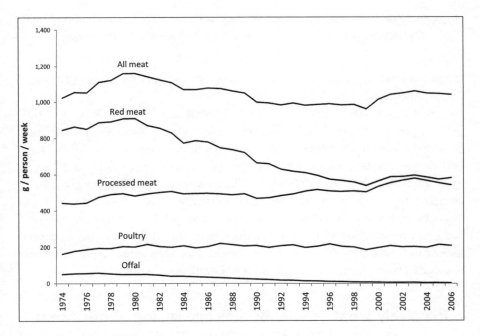

Source: Allender et al (2008)

Figure 12.1 Consumption of red meat, poultry and processed meat in the United Kingdom, 1972–2006

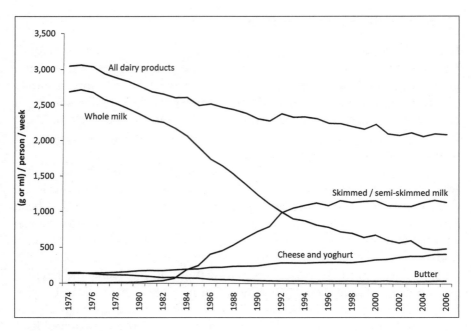

Source: Allender et al (2008)

Figure 12.2 Consumption of whole milk, semi-skimmed milk, skimmed milk and processed dairy products in the United Kingdom, 1972–2006

led to a decline in saturated fat intake from 19 per cent to 14 per cent since 1975 (Allender et al, 2008).

It is difficult to quantify exactly the contribution of these changes in MDP consumption to the decline in CVD mortality in the UK but the contribution is likely to have been considerable. Changes in consumption of MDPs have had similar effects on CVD rates in other countries such as Finland, Norway, the Czech Republic and Poland. In Finland, where it has been possible to compare the contribution of both improvements in treatment for CHD and changes in diet and other lifestyle factors that affect the incidence of CHD, it has been shown that 33 per cent of the decline in CHD mortality observed there can be attributed to a decline in blood cholesterol levels, in turn largely associated with changes in MDP intake (Laatikainen et al, 2005)

Precisely what official bodies are saying about meat and dairy consumption

Despite the changing consensus around the health consequences of MDP consumption it is only comparatively recently that official government departments and agencies have begun to suggest that MDP consumption is anything other than good for health. Moreover advice about MDP consumption has tended to be about switching from those products that are very high in saturated fat or salt to those that are less high, rather than to cut down on MDPs themselves.

However, very recently it has begun to be suggested that a reduction in MDP consumption might be advisable. This has generally been in the context of an examination of what constitutes a diet that would be good for the environment as well as health. So for example the Swedish Government recently recommended that people should eat less meat for environmental reasons noting that 'From a health perspective, there is also no reason to eat as much meat as we do today' (National Food Administration and Swedish Environmental Protection Agency, 2009).

In the UK the Sustainable Development Commission (SDC), in a report that aimed to define both a healthy and environmentally sustainable diet, said that '[c]hanges likely to have the most significant and immediate impact on making our diets more sustainable, in which health, environmental, economic and social impacts are more likely to complement each other' included '[r]educing consumption of meat and dairy products' (SDC, 2009) but it should be noted that the SDC merely advises the UK Government and is not the UK Government itself.

At around the same time members of an international collaboration of scientists supported by a consortium of funding bodies coordinated by the Wellcome Trust (whose findings were recently published in the *Lancet*) suggested that there would be considerable health benefits from food and agriculture strategies aimed at reducing greenhouse gas emissions (Friel et al, 2009). Margaret Chan, Director General of the World Health Organization (WHO), commenting on the article stated that 'reduced consumption of animal products in developed countries would bring public health benefits' (Chan, 2009).

However, the WHO has never said this in its official reports. The definitive WHO statement on diet and health is *Diet, Nutrition and the Prevention of Chronic Diseases* (WHO, 2003) and this does not recommend a reduced intake of animal products within developed countries but merely notes that: 'Excessive consumption of animal products in some countries and social classes can lead to excessive intakes of fat.'

In the UK the most authoritative expert body on nutrition is the government's Scientific Advisory Committee on Nutrition (SACN). SACN does not advise a reduction in meat and/or dairy consumption in the UK. The closest it gets is to say: 'Lower consumption of red and processed meat would probably reduce the risk of colorectal cancer. Although the evidence is not conclusive, as a precaution, it may be advisable for intakes of red and processed meat not to increase above the current average (70g/day) and for high consumers of red and processed meat (100g/day or more) to reduce their intakes' (SACN, 2009).

Moreover SACN have never argued that we need to cut down on MDPs themselves to reduce the risk of CVD. They have merely indicated a need to cut down on MDPs high in fat, particularly saturated fat (and also salt) (Department of Health, 1994).

Finally the UK Food Standards Agency – responsible for turning SACN's advice into dietary recommendations for the general public – have never suggested that we need to cut down on MDPs but merely to switch to MDPs products that are lower in saturated fat and salt. The most important pieces of consumer advice that the UK Food Standards Agency publishes is the Eatwell Plate (Food Standards Agency, n.d. 2) (originally developed in the 1990s and then called the 'National Food Guide', Hunt et al, 1995). This is a pie chart in the form of a plate and it represents the amount of foods that consumers should eat from each of five food groups: 'fruit and vegetables', 'bread, potatoes and other cereals', 'meat, fish and alternatives', 'dairy products' and 'fatty and sugary foods'. The Eatwell Plate indicates that while we need to eat more 'fruit and vegetables' and 'bread, potatoes and other cereals' and fewer 'fatty and sugary' foods, we are eating about the right amount of 'meat, fish and alternatives' and 'dairy products'.

Conclusion

How then might we decide: 'How much meat and dairy is optimal for health?' and 'What type of meat and dairy products are optimal for health?' These are different, separate, but of course interrelated questions.

How much meat and dairy is optimal for health?

Our discussion above about the positive and negative aspects of MDP consumption suggests that the costs and benefits to health from MDP consumption can be balanced against each other, and a consensus can therefore be achieved as to the exact level of meat and dairy consumption that should be recommended in order to gain optimal benefits with acceptable risk levels. In reality, this balancing of the positive and negative aspects is very complicated. There are two main reasons for this.

First, human beings need to eat and drink to live, and MDPs represent only a proportion of the total number of foods and drinks that are available to meet energy and nutritional needs. The consumption of MDPs cannot be considered a closed system in which an adaptation of consumption levels will have negligible impacts on other behaviours – rather, if consumption of MDPs is reduced then consumption of other foods must increase, and vice versa. The nature of this displacement effect is not well understood but could have a profound impact on the health effects of reducing or increasing MDP consumption. For example, successful vegans generally tend to have lower adverse health outcomes compared to the general population, which suggests that it is possible for some people to remove all MDPs from the diet and replace the positive nutritional aspects of meat and dairy with nutrients from plant-based sources, fortified foods or dietary supplements. However, not all people who remove MDPs from their diets are successful at this replacement, and some revert to the consumption of MDPs in order to improve their health.

There is another important aspect of this displacement effect: it may obscure some of the mechanisms for the relationship between MDP consumption and health. Adverse health outcomes associated with high MDP intake may in part be due to a related low intake of fruit and vegetables, for example, but adjustment for this potential confounding in observational studies is hindered by the difficulty of accurately measuring dietary intakes.

The second reason for the difficulty in balancing the positive and negative aspects of MDP consumption is that the positive health aspects tend to be about averting poor quality of life in the immediate term (such as averting diarrhoea or dermatitis), whereas the negative aspects tend to be about increased risk of mortality in the long term, and these two metrics are notoriously difficult to combine and compare. Methods have been developed that combine years of life lost due to early death and years of life lost in ill health within a single index (e.g. the disability adjusted life years lost (DALY)). But DALYs (and similar indices) have yet to be used to compare the positive and negative aspects of MDP consumption.

Not only must such an index compare the debilitating effects of ill health with mortality, it must also compare immediate health status with increased risk of disease in the future. Unsurprisingly, most research regarding the impact of diet on health is conducted using more conventional health outcomes – incidence of disease, mortality etc. – and even then it is rare for studies simultaneously to consider both disorders due to nutrient deficiencies and chronic disease as health outcomes, since in the developed world CVD and cancer is far more prevalent than disorders due to nutrient deficiencies. This is reflected in the analyses that have specifically looked at the health impacts of consumption of MDPs in the diet, for example in the meta-analysis of studies that have assessed the effects of consuming MDPs on the incidence of cancer from specific sites carried out by the WCRF mentioned above (WCRF/AICR, 2007) and more recently in the results of a much publicized prospective study of diet and health in over half a million people (Sinha et al, 2009).

This all being said, it does seem likely that the negative consequences of eating large amounts of meat (and to a lesser extent dairy) as in the UK might outweigh the positive benefits. It is certainly time to see if this analysis can be done and to revise government advice (e.g. in the form of the Eatwell Plate) accordingly.

What type of meat and dairy products are optimal for health?

On the face of it this seems a simpler question. It seems clear that if we are to eat MDPs then on a routine basis we should eat those products that are lower in saturated fat and salt, e.g. fresh meat rather than processed meat, semi-skimmed and skimmed milks rather than whole milk, lower fat, lower salt cheeses, low fat, preferably polyunsaturated spreads rather than butter, etc. Furthermore, on the basis that it seems that red and processed meats eaten in excess, rather than white and unprocessed meats, increase the risk of colorectal cancer we might consider eating less of the former type of product and more of the latter (particularly if we are high meat consumers). It may also be worth paying some attention to new ideas about how grass-fed animals produce healthier MDPs than intensively reared animals.

But whatever the answer to the question 'What type of MDPs are optimal for health?' the more important question is surely 'How much MDP is optimal for health?'. This is one of the questions we need to answer if we are to save the planet.

References

Allender, S., Peto, V., Scarborough, P., Kaur, A. and Rayner, M. (2008) 'Coronary heart disease statistics 2008', British Heart Foundation, London

Cannon, G. (2003) *The Fate of Nations: Food and Nutrition Policy in the New World*, The Caroline Walker Trust, London

Chan, M. (2009) 'Cutting carbon, improving health', *The Lancet*, vol 374, no 9705, pp1870–1871

Department of Health (1991) *Dietary Reference Values for Food Energy and Nutrients for the United Kingdom, Report on Health and Social Subjects 41*, HMSO, London

Department of Health (1994) 'Nutritional aspects of cardiovascular disease', Report on Health and Social Subjects 46, HMSO, London

Fiddes, N. (1991) *Meat: A Natural Symbol*, Routledge, London

Food Standards Agency (n.d. 1) 'Nutrition essentials', available at www.eatwell.gov.uk/healthydiet/nutritionessentials, last accessed 28 February 2010

Food Standards Agency (n.d. 2) 'The Eatwell Plate', available at www.eatwell.gov.uk/healthydiet/eatwellplate, last accessed 26 February 2010

Food Standards Agency and Department of Health (2004) 'The national diet and nutrition survey: Adults aged 19 to 64 years (2000/2001)', FSA and DH, London

Food Standards Agency and Department of Health (2010) 'National diet and nutrition survey. Headline results from year 1 of the rolling programme (2008/2009)', FSA and DH, London

Friel, S., Dangour, A., Garnett, T., Lock, K., Chalabi, Z., Roberts, I., Butler, A., Butler, C., Waage, J., McMichael A. J. and Haines, A. (2009) 'Public health benefits of strategies to reduce greenhouse-gas emissions: Food and agriculture', *Lancet*, vol 374, no 9706, pp145–152

Givens, D. I. (2005) 'The role of animal nutrition in improving the nutritive value of animal derived foods in relation to chronic disease', *Proceedings of the Nutrition Society*, vol 64, pp395–402

Hunt, P., Rayner, M. and Gatenby, S. J. (1995) 'A national food guide for the UK? Background and development', *Journal of Human Nutrition and Dietetics*, vol 8, pp315–322

Key, T. J., Fraser, G. E., Thorogood, M., Appleby, P. N., Beral, V., Reeves, G., Burr, M. I., Chang-Claude, J., Frentzel-Beyme, R., Kuzma, J. W., Mann, J. and McPherson, K. (1999) 'Mortality in vegetarians and nonvegetarians: Detailed finings from a collaborative analysis of 5 prospective studies', *American Journal of Clinical Nutrition*, vol 70, pp516S–524S

Laatikainen, T., Critchley, J., Vartianen, E., Saloma, V., Ketonen, M. and Capwell, S. (2005) 'Explaining the decline in coronary heart disease mortality in Finland between 1982 and 1997', *American Journal of Epidemiology*, vol 162, pp764–773

Mann, J. (2009) 'Vegetarian diets. Health benefits are not necessarily unique but there may be ecological advantages', *BMJ*, vol 339, pp525–526

McKee, F. (1997) 'The popularisation of milk as beverage during the 1930s', in Smith, D. E. (ed.) *Nutrition in Britain*, Routledge, London, pp123–139

Micha, R., Wallace, S. K. and Mozaffarian, D. (2010) 'Red and processed meat consumption and risk of incident coronary heart disease, stroke and diabetes mellitus. A systematic review and meta-analysis', *Circulation*, vol 121, pp2271–2283

Millward, D. J. and Garnett, T. (2010) *Food and the Planet: Nutritional Dilemmas of Greenhouse Gas Emission Reductions through Reduced Intakes of Meat and Dairy Foods*, Proceedings of the Nutrition Society, vol 69, pp103–118, Cambridge University Press, doi: 10.1017/S0029665109991868

Mozaffarian, D., Katan, M.B., Ascherio, A., Stampfer, M. J. and Willett, C. (2006) 'Trans fatty acid and cardiovascular disease', *New England Journal of Medicine*, vol 354, pp1601–1613

National Food Administration and Swedish Environmental Protection Agency (2009) *The National Food Administration's Environmentally Effective Food Choices: Proposal Notified to the EU*, NFA, Stockholm

Scarborough, P., Nnoaham, K., Clarke, D., Capewell, S. and Rayner, M. (in preparation) *Modelling the Impact of a Healthy UK Diet on Cardiovascular Disease and Cancer Mortality*

SACN (2009) *Iron and Health*, SACN, London

Sinha, R., Cross, A. J., Graubard, B. I., Leitzmann, M. F. and Schatzkin, A. (2009) 'Meat intake and mortality: A prospective study of over half a million people', *Archives of Internal Medicine*, vol 169, pp562–571

Skeaff, C. M. and Miller, J. (2009) 'Dietary fat and coronary heart diseases: Summary of evidence from prospective cohort and randomised controlled trials', *Annals of Nutrition and Metabolism*, vol 55, pp173–201

Strazzullo, P., D'Elia, L., Kandala, N. B. and Cappuccio, F. P. (2009) 'Salt intake, stroke and cardiovascular disease: Meta-analysis of prospective studies', *BMJ*, vol 339, b4567

Sustainable Development Commission (2009) 'Setting the table: Advice to government on priority elements of sustainable diets', Sustainable Development Commission, London

Vegetarian Society (n.d.) 'History of vegetarianism', available at www.vegsoc.org/info/developm.html, last accessed 26 February 2010

Von Liebig, J. (1846) *Animal Chemistry*, Scholarly Publishing Office, University of Michigan Library; 3 Revised edition (13 September 2006), available at http://tinyurl.com/397tgua, last accessed 25 May 2010

Von Liebig, J. (1847) *Researches on the Chemistry of Food*, Taylor and Walton, London, available at http://tinyurl.com/37ev953, last accessed 24 May 2010

Webster, C. (1997) 'Government policy on school meals and welfare foods 1939–1970', in Smith, D. E. (ed.) *Nutrition in Britain*, Routledge, London, pp123–139

WCRF/AICR (1997) 'Food, nutrition, physical activity and the prevention of cancer: A global perspective', World Cancer Research Fund/American Institute for Cancer Research, Washington DC

WCRF (2007) 'Food, nutrition, physical activity and the prevention of cancer: A global perspective', World Cancer Research Fund/American Institute for Cancer Research, Washington DC

WHO (2003) *Diet, Nutrition and the Prevention of Chronic Diseases*, WHO Technical Report Series, No 916, WHO, Geneva

WHO (2007) 'Protein and amino acid requirements in human nutrition', report of a joint WHO/FAO/UNU expert consultation, WHO Technical Report Series, No 724, WHO, Geneva

Part 4
Ethical and Religious Approaches to Animal Foods

13

Developing Ethical, Sustainable and Compassionate Food Policies

Kate Rawles

Introduction

The anthropologist nervously stroked her antennae. The renewal – or not – of her research grant most likely depended on what she said next.

'And the update on Earth food production systems?'

The request came from the chief Martian Institute Research Coordinator himself. This was it.

'Well', she began, 'we have observed very different patterns of food consumption emerging over the last hundred or so Earth years. Some sectors of Earth human population are especially intriguing, and it is these sectors I believe warrant further study.'

'Go on', the chief said.

The anthropologist flicked through her files.

'We have observed', she said, hesitantly, 'a very substantial rise in the consumption of "meat" – flesh from deceased animals – across a range of human populations, largely facilitated by changes in farming technology. What many of these Earth beings have now realized, however', she said, warming to her theme, 'is that this rise in meat consumption, initially thought to be beneficial, is generating multiple interlocking problems. These include human health issues, such as ischaemic heart disease, obesity and possibly cancers as well as aggravated incidence of zoonotic diseases, such as swine flu. In addition, meat production as practiced in the societies under observation is highly water intensive and, typically, extremely inefficient in relation to land use, thus exacerbating pressure on critical resources at a time when both human populations and resource scarcity are increasing. The systems in question are also extremely negative in relation to the welfare of the beings who are "produced". And, perhaps most significantly, recent research by Earth scientists themselves has revealed that the production of meat, especially from ruminants, is one of the largest single contributors to anthropogenic climate change. This climate change is on a trajectory predicted to cause the extinction of about one half of Earth's current suite of species and widespread resource conflict, death and displacement of the human species, by the end of the Earth century. Many negative impacts will of course occur well in advance of that and a significant number are already underway. In our view, as well as that of Earth scientists, it is not clear that Earth's human societies can survive this degree of climate change. They certainly cannot in their current form.'

'Fascinating, indeed', the chief MIRC acknowledged. 'Presumably your research interest lies in assessing the efficacy of the humans' strategies for reversing this situation?'

'No', the anthropologist replied. 'That is precisely the point. It lies in our discovery – quite unique in my cosmic career – that this species appears intent on acting against its own interests. All these problems could be greatly reduced, if not resolved, by eating non-animal derivatives – without loss of nutritional content. Yet, despite humanity's rapidly burgeoning knowledge and understanding of these issues, and notwithstanding their reasonably well-developed analytical brains (by outer arm Milky Way standards at least) and even an average compassion rating (ironically especially well developed in relation to large animals with eyes), dominant sectors of these societies are not only maintaining but even increasing their own meat consumption, as well as exporting it to previously low-meat consumption sectors of the Earth community as a desirable norm.'

'Cosmos above!' the chief MIRC said, visibly astonished. 'What, in your view, is the likelihood of this policy – if it can be called a policy – being reversed?'

'That greatly depends, in our view, on the extent to which the current and highly fortuitous Earth economic "credit crunch" continues to allow a rare window of opportunity for intervention in the typically rather blind allegiance to a largely market-led system by principled, compassionate, far-sighted, scientifically informed ethicists creating and taking up positions of influence within a greatly strengthened International Food Policy decision making board.'

'Unlikely, in other words', said the chief MIRC. 'I see. Grant application request extended – we can clearly learn more about the features of fatally flawed decision-making processes from this research. Meanwhile, memo to security: advise in the strongest terms that we continue to reprogramme all satellite imagery in order to foil attempts by Earth scientists to find our community here on Mars. Record of peaceful interactions by Earth beings with benign cultures, particularly under times of resource stress, is exceedingly poor. Next applicant!'

***

The first step in developing ethical, sustainable and compassionate food policies is to acknowledge that we need them. In the 2006 Stern Report, Sir (now Lord) Nicholas Stern famously described anthropogenic climate change as the greatest market failure ever witnessed. Climate change, he argued, can be seen as the disastrous but unintended outcome of millions of consumer choices made in a relatively unregulated market shaped largely by short-term economic priorities – in the wider context, other analysts might add, of a pro-consumption, pro-growth socio-economic world view. Stern is not a critic of pro-growth world views but he does argue strongly that tackling climate change will require systematic and rigorous intervention in market-led systems, in the light of a very different set of priorities.

Stern might have made similar points about high levels of meat consumption had he been studying it and had he had the good fortune to view the issue with the clarity and lack of bias of a detached Martian observer, as opposed to the somewhat distorted perspective occasionally associated with the vested interests of, for example, a meat eater resident on Earth. The outcome of millions of consumer choices, made in a socio-economic context in which, for various reasons and in various ways, meat eating is encouraged, advertised and advocated, has lead to a situation in which meat production and consumption in its current form is a major contributor to a suite of serious problems, as sketched above and explored throughout this book. Nobody has deliberately set out to create this system but this is the system we have. The idea that ethical, sane approaches to food production will spontaneously emerge from it – from the 'free' market – with a bit of consumer education and product labelling is, unfortunately, delusional. If we are to reorient food production along sane, sustainable, compassionate and ethical lines then we need to intervene. In what follows below, I will sketch out the principles and values that should, arguably, guide this intervention.

Why animals?

But first, why should we be concerned about compassion and ethics in relation to animals? Implementing actions and policies to deal with climate change and other urgent environmental issues is – at least in theory! – relatively easy to justify. Preserving our own habitat is clearly in our own interests, even if it has taken us a while to figure this out (and not that this conclusion is exactly dominating mainstream political activity.) The animal welfare agenda, however, can be seen as irrelevant to our interests. This agenda attempts to implement the ethical obligations that we have towards *other* sentient beings; to make compassion for other animals systematic and practical, with positive implications for how many millions of real animals live and die. On the face of it, however, whether or not farm animals, for example, live out their lives in poor conditions, does not appear to affect our own human lives very much at all. Add to this the ever-mounting concern about climate change, and the priority this issue is at last beginning to receive, as well as the more recent recession-related economic agenda and it is easy to see how concern with animal welfare can be marginalized. It is not unusual, now, to hear the argument that animal welfare is something we will have to trade off against more urgent concerns; a luxury that, in the current economic and environmental context, we will have to relinquish as simply unaffordable.

In the rest of this chapter, I will argue that this is deeply mistaken. High standards of animal welfare are – or should be – a core value of food production and a critical, if often neglected, dimension of sustainable development more generally. Indeed, far from being a luxury, I will argue that dealing with poor animal welfare, or at least the mindset that legitimizes it, is a necessity. For it is precisely this mindset that has given us climate change *and* the recession. And, to paraphrase that much used Einstein quote, 'you can't solve a problem with the same kind of thinking that caused it'.

The triple crunch context

The recession, and the 'credit crunch' that initiated it, have dominated the news and taken centre stage in political activity from local to global. As the main focus of the G20, our current clutch of world leaders have rightly grappled with the implications of a global recession, not least in relation to what rising unemployment, rising inequity and a slowdown in economic growth actually mean in terms of human lives and human suffering. But a number of analysts have argued compellingly that the credit crunch will be as nothing compared to the impacts of climate change or the so-called 'nature crunch' more generally.

In 2006, Stern published the report on climate change mentioned above, in which he essentially overturned the last argument against robust action on climate change that had any shred of credibility. He did this by demonstrating that the costs of letting climate change rip and mopping up the damage would vastly outweigh the costs of taking action now to try to mitigate its worst effects. Stern has just published a new book in which he points out that the scientific evidence on which those conclusions were based has been revealed as wildly optimistic. Climate change predictions in relation to temperature and sea level rise, and their likely impacts on social and ecological systems, are systematically worsening. In 2006, Stern was optimistic about our chances of keeping the average global temperature increase below two degrees above pre-industrial levels. He now thinks we are heading almost inexorably for four degrees by the end of the century with much of the increase happening sooner rather than later. Two degrees is the upper limit before 'runaway' climate change becomes an outcome with high probability. At four degrees, the estimates are that half – half! – of all wild species will become extinct and about half the Earth's land surface will become desert. In conjunction with widespread water shortage, especially in areas where water is sourced from glacial melt, massive displacement of people and resource conflict is inevitable, to say nothing of the impacts on other species. I heard Lord Stern talk in London a few weeks before writing this and, while the dawning reality of climate change has long been appalling in relation to its impact on others, it is the first time I have actually felt afraid for myself.

At the same time, human encroachment on habitat, not least for agriculture, is amongst the contributors to what has been called the sixth great extinction, with a recent United Nations report stating that 'nature is not just a luxury some of us enjoy at the weekends' but critical to our own survival as a species. According to the United Nations, our mid- to long-term survival is now genuinely in doubt because of the impact of the way we acquire our resources on biodiversity and, in the shorter term, this same impact has meant that the aims of reducing malnutrition and other aspects of absolute poverty as set out, for example, in the Millennium Development Goals, have become all but unattainable. As Stern and others show, the credit crunch is indeed nothing to the nature crunch, even if our politicians and the media might have it the other way around.

In short, considering the increases in poverty and inequality both locally and globally that these combined economic and environmental issues are delivering right now, and a cluster of recent research saying that inequality *per se* (in other words, independently of the absolute position of the worst off) has significant negative impacts on human health and well-being – as well as some truly appalling figures just released that show climate change alone is currently responsible for 300,000 deaths a year with 500 million people classified as at extreme risk from climate change weather-related events – then it makes sense to talk about a triple crunch: economic, environmental and social. And that, of course, simply emphasizes yet again that the need for humans to get their collective act together in relation to sustainable development – often characterized in terms of the sustainability diamond, with economy, environment and society as its three points (see below) – is absolutely compelling.

Animal welfare should be a core value – two arguments

Why, then, given the above as an outline of the context, should animal welfare be considered a core value, and one that – alongside environmental, climate and human health and well-being concerns – should guide the ethically appropriate recrafting of our food policies? I will sketch two arguments, the first to do with sustainable development and the second to do with problematic mindsets.

Animal welfare and sustainable development

To paraphrase something that Roland Bonney (a Director of the Food Animal Initiative, Oxford) once said: sustainable development is about being able to do tomorrow what you can do today. Fine as far it goes, as he pointed out. But the fact that today's societies, complete with gross inequity and unacceptable levels of poverty, could certainly be around tomorrow – a little under one in five people across the world currently lack clean drinking water and basic nourishment – clearly shows that sustainable development is not just about 'carrying on'. It is about carrying on in a way that manifests our core values as civilized societies. Or, remembering Gandhi's famous quip, 'what do I think about Western civilization? I think it would be a good idea', sustainable development is *ethically aspirational*. It is about the values we *want* societies to manifest; the values we consider critical if we are to continue our residence on Earth in civilized, humane and meaningful ways.

The core values of sustainable development are often represented as points on a triangle – or legs on a three-legged stool.

What is striking from the perspective of this chapter is that, as I have argued elsewhere, concern with animal welfare doesn't really fit at any point. Economic development is (or is supposed to be) in the interests of meeting human needs and improving human quality of life. The 'society' end of the triangle represents values to do with equity, within and between generations, and also the opportunity for meaningful participation in social

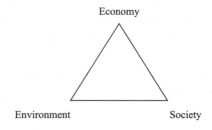

Figure 13.1 The sustainable development triangle

and political institutions – again between humans. And 'environment' is concerned with protecting resources for people or, on less anthropocentric interpretations, coexisting with other species and ecosystems, valued in their own right as well as in relation to their undeniable usefulness to ourselves. On either interpretation, the environmental focus is on 'ecological collectives' of various kinds. But the concern with animal welfare is a concern with animals as individuals rather than as species or part of ecosystems; and it is a concern with animals that are typically domesticated or tamed rather than wild. Animal welfare does not naturally fit any point of the sustainable development triangle and, although this may be changing, it has often been excluded from mainstream sustainable development rhetoric, concerns and policies.

We've already argued that sustainable development is ethically aspirational. Indeed, with its focus on dealing with poverty and intergenerational equity, sustainable development clearly has ethics at its very core. But there is a whole flock of arguments to the effect that restricting ethics exclusively to inter-human concerns makes no rational sense, ranging from Peter Singer's condemnation of 'speciesism' to Mary Midgley's 'animals are part of our community' approach.[1] I can't review these here but suffice to say that, if sustainable development is ethically aspirational (and, if it isn't, it is of highly dubious value) then the humane treatment of the animals we use for our own purposes and that we are responsible for in relation to farming, pet-keeping and so on must surely be included. What counts as 'humane' is then another debate, of course. But if there is a non-speciesist argument to the effect that humane treatment of animals in our care is irrelevant, I'd be really interested to read it. Meanwhile, I argue that we should adapt the sustainable development triangle to a diamond, with animal welfare as the fourth point. Without it, a key ethical commitment, and one that affects billions of sentient beings, can simply be overlooked.[2]

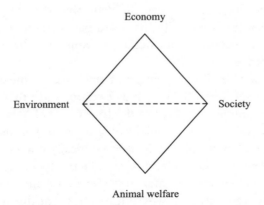

Figure 13.2 The sustainable development diamond

Of course, there is a further ethical issue here about the legitimacy of both domestication and animal-based agriculture. Tom Regan, for example, maintains that campaigning for improved farm animal welfare is as ethically misguided as working to improve the well-being of slaves.[3] In both cases, he believes, the institution should be abolished. Doing justice to this argument is beyond the scope of this chapter and the abolition of domestication in general and farmed animals in particular is clearly a long way off. Moreover, even if Regan were right that this should be our ultimate aim, it certainly wouldn't follow that poor animal welfare in existing (pre-abolition) farms would somehow be acceptable in the interim.

Instrumental thinking underpins the triple crunches – and the dismissal of animal welfare as a luxury

Various philosophers and other commentators have discussed 'instrumental rationality' as part of the mindset that has its roots in Enlightenment thinking and that has proved to be, to say the least, a double-edged sword. On the plus side, it has allowed us to develop fantastically efficient systems and greatly increased production, in agriculture amongst other contexts. But the negatives are stacking up to the point of being overwhelming.

Instrumental rationality or 'means/end' thinking focuses on one particular value, or suite of values, most commonly economic growth and/or profit, often in the short term. It then looks at the world, and everything in it, entirely from this perspective. Thus the environment is viewed as a set of resources to be managed for human gain rather than as a complex, living system on which we utterly depend and of which we are a part; people are viewed as workers or resources of other kinds, or irrelevant; farm animals are viewed as cogs in a machine designed with the aim of turning animal fodder into human food as efficiently as possible. From this perspective, more or less anything becomes acceptable in the efficient pursuit of profit or economic growth (to stay with that example) and other values – such as environmental protection, social welfare or animal welfare – are *all* overridden as luxuries we can't afford.

A current example of concern about this kind of means/end thinking is the fear expressed by NGOs such as Oxfam, and other commentators, that the recently increased power of the International Monetary Fund will only exacerbate current levels of poverty and inequity in developing countries, as the kind of financial 'aid' it offers typically comes with strings – if not steel hawsers – attached. These often take the form of requirements to cut back on social welfare, in order to achieve efficiency in relation to the goal of increased economic growth: but in ways that drastically impact on the poorest. More generally, a key feature of instrumental thinking in this sort of context is that a whole host of beings are considered, well, instrumentally – as a means to the end of economic growth in ways that are typically detrimental to their own well-being. The neoliberal model of unfettered global markets is, of course, a classic example of instrumental rationality. The justification usually offered in its defence – until recently almost unchallengeable without being labelled a lunatic – is that only with increasingly unfettered markets can we achieve continuous economic growth, and only with continuing economic growth can we tackle poverty (despite the mass of evidence that, decades of growth notwithstanding, the poorest have remained poor and their number has increased) or, even more bizarrely, that only with economic growth can we afford to fix the environment.

This kind of thinking has its place. As indicated above, it has facilitated, for example, increased productivity across a range of contexts. But when it becomes dominant, in the way that it has in modern industrialized societies, the result is a focus on one set of values that are too limited and that are pursued at the expense of a myriad other values of critical importance. This kind of means/end, instrumental thinking has delivered us economic systems that have become increasingly divorced from 'real wealth' as well as from the values such as trust, openness, honesty, accountability and caring that turn out to be crucial to the sustainable functioning of the economic system itself. This kind of thinking has overridden the need to look after the ecological systems we are part of to the extent that climate change and biodiversity loss now threaten our own survival as a species – let alone that of millions of the other species we coexist with. It has led us to ignore the inconvenient truth that our current model of infinite growth is totally at odds with the reality of living on a planet that has ecological limits – limits to space, to resources, to the capacity to absorb pollution. And it has given us the kind of intensive systems that systematically deny animals basic welfare as outlined, for example, in the Five Freedoms (see below).

If we are to deal with the nature crunch, the credit crunch, the poverty crunch and the welfare crunch – if we are to move towards sustainable development in any meaningful sense of that term – we absolutely have to challenge this kind of thinking, not legitimize it in yet another guise. The view that animal welfare is a luxury we can't afford and should be overridden to achieve economic and environmental priorities is precisely a case in point: a clear example of means/end, instrumental thinking, with appalling practical consequences for the billions of animals that go through our farming systems every year. Instead of seeking to ease our economic and environmental problems by compromising on animal welfare – a bit like responding to the knowledge that a loud noise is making us deaf by turning the sound up – we need to tackle the problematic thinking that underpins all of these problems, in the full understanding that they are not separate, but profoundly interconnected, issues.

A wiser world view

What, then, is the way forward? First, we need to insist that our food production systems be shaped by other values beyond efficiency in profit production. The way we produce our food is not just another industry. Food production is critical to our health and well-being, and the systems we put in place to produce our food have immense impacts upon the health and well-being of the animals involved, as well as on the wider environment – on other living beings, species, ecological systems. Human health, environmental well-being, good animal welfare – these are all important values that our systems for providing us with food need to embody and reflect, not displace in the name of efficiency. Food production should not be – *cannot* be if we are to continue our tenancy on planet Earth – just a question of efficiency. It plays a critical role in sustainable development and it urgently needs to be reoriented along explicitly ethical lines.

The second point, as Stern argued in the climate change context, is that we should be absolutely clear that we can't expect ethical and sustainable food production systems to be delivered, as if by magic, by the market. The market cannot deliver unaided in this way, for reasons we've outlined above. We need to intervene; to intentionally reorient our food production systems in relation to a much richer and more appropriate set of values than it currently serves, in the context of a saner, wiser world view.

What would such a worldview entail? We can develop this by looking briefly at the problematic one in a little more detail. Scratch the surface of modern industrialized societies and, alongside the predominance of means/ end thinking, you find a whole set of attitudes, beliefs and assumptions that add up to an implicit world view. Once this world view is made explicit, our stunningly unsustainable approach to meeting our needs suddenly appears much less surprising. Indeed it is hard to see how societies with this kind of conceptual underpinning could be anything *other* than unsustainable.

Four interconnected aspects of this world view merit particular comment in this context. First, as we've seen, this is a world view that typically understands the Earth and everything in it as a vast repository of resources for the sole benefit of humans. In relation to the other-than-human world, instrumental thinking is supreme. Second, these resources have often been assumed to be infinite. Take these together and, in Ray Anderson's words, you get a picture of the world 'as a sort of gigantic production system, capable of producing ever-increasing outputs'. Third comes the particularly bizarre phenomena that I've referred to as the 'allotment mindset'. This is the belief that humans are somehow *outside* ecological systems. Ecological systems – the 'environment' in general – are out there, like a vast allotment, and we stride out and take from it when we need to. But we are not really *in* it. An extraordinary techno-optimism often accompanies this belief, to the effect that, even if Earth's resources turn out not to be infinite after all, we don't need to worry because we will find ways of manufacturing our own resources. So long, Earth, and thanks for all the fish – but now we can make our own. Fourth, and clearly rooted in the previous aspects of the modern, industrial world view, is the view that not only is increasing consumption of resources (by increasing numbers of people) unproblematic; it is to be encouraged. If we accept this world view, then growth of consumption leads to growth of economy and economic growth is good. Indeed it is, as we have seen, our overriding aim and number one value. It is certainly not a problem. Economic growth, premised on continuously increased consumption, can and should continue indefinitely; it is the best and only way of ensuring high standards of living for all.

The outline characteristics of a saner, wiser world view, and the ethics that would follow from it, will by now be obvious. For a start, we need to challenge our own out-of-control human-centrism. Of course the environment is a resource for humans and, like all species, we have to relate to it partly in this way. But it is not only a resource. The vast complex of astonishing diversity, energy and sheer will to live that is 'the environment' has value far beyond its usefulness to us. Basking sharks and blue tits, savannah and rainforests, clouds, stars and streams – as well as cats, dogs, cattle and pigs – have value beyond the extent to which one species amongst millions happens to need them. To deny this is to take an astonishingly arrogant stance, positioning humans as the only species of true worth and the rest of relevance only in relation to ourselves. This is a pre-Copernican view of ethics; the values equivalent of believing that the sun spins around the Earth.

In addition, we need to acknowledge and fully understand that we are part of ecological systems, not apart from them. We are not on the outside looking in. Our experience of life may distance us from the reality that all our basic needs are sourced in planet Earth and its ecological and physical systems. But however many layers of technical brilliance intervene between natural resources and our end products, we cannot detach ourselves from our ultimate dependence on Earth ecology. For all our technology, we remain earthbound creatures, relying on ecological systems for our basic needs, including food, as the reports of the Millennium Ecosystem Assessment and the Martian anthropologists have both made clear. Finally, and crucially, we have to acknowledge that Earth's biophysical systems have limits. We cannot, as modern societies have assumed we can – and astonishingly some forms of farming have to be included here – simply ignore biophysical realities and endlessly extract resources at one end and produce pollution at the other without consequence. We need to reorient the way we live and meet our needs so that we can live within biophysical limits, not endeavour to ignore them.

This is where I part company with Stern. His analysis of the climate change problem is compelling but he only has part of the solution. Stern believes we can simply (well, not simply, but we can) decouple the economy from carbon and continue to grow. Continue with the same mindset, the same instrumentalizing view of the Earth and other living things (including people); the same commitment to growth that, even if it acknowledges limits to Earth's capacity to absorb carbon fails to acknowledge there are limits to space; to the Earth's capacity to absorb other forms of pollution; to ecological systems' ability to continue to function even as we decimate the species that make them up. In sum, not only is the challenge of decoupling the economy from carbon actually, in the context of increasing consumption and increasing human population, immense beyond Stern's recognition; the Earth has other limits too. We cannot grow indefinitely on a finite planet. It is already the case that, as World Wildlife Fund (WWF) has so powerfully put it, if everyone on earth lived the lifestyle of an average western European, we would need three planet Earths. Yet this is the lifestyle whose consumer-oriented material values underpin what is meant by 'development', 'success' and 'progress' and are exported around the world as something to be emulated. The production of meat, in industrialized countries, thus needs to be understood as embedded in agricultural systems that are profoundly unsustainable; that are themselves embedded in industrialized cultures whose concepts of progress and success are profoundly unsustainable and that have set human societies on a trajectory that is, literally, self-destructive.

Crucially, then, we need a world view that supports a different set of values. Values that emphasize quality of life rather than a materialistic high standard of living; and values that acknowledge the worth of people, animals and environment in their own right, not just as a set of resources to service out of control industrialized growth economies. We need to get a grip of our runaway market-led system that exemplifies instrumental, means/end thinking, that is embedded in a deeply flawed world view and that has delivered us the range of interconnected problems sketched above. We need to reshape the current system, along firmly ethical, values-led lines, in the context of a wiser, saner, world view outlined above – fully realizing our own interdependence with other living things, understanding ourselves as part of a wider ecological and community (and one that includes our domesticated animals) rather than as detached managers of a set of resources and aspiring towards a very different conception of quality of life than the current materialistically conceived understanding of a what 'high' standard of living entails.

An outline ethical framework

To translate this wiser, saner, compassionate world view into actionable ethical policies, I suggest we need a quadruple bottom line approach – in keeping with the sustainability diamond.

The quadruple bottom line would make positive recommendations and some core prohibitions. Policies would be judged against their positive and negative contributions to:

- Environmental values – with a range of indicators here in relation to biodiversity, ecosystem integrity, climate change, other forms of pollution reduction and so on.
- Social values – with a focus on reducing inequality across current generations; increasing participation; and developing less materialistic, conceptions of quality of life (which could also be more fulfilling).
- Economic values – with a refocusing on the core aim of meeting human needs both now and in the future (and a big issue here in relation to distinguishing real needs from wants or luxuries).
- Animal welfare values – articulated, for example, in relation to the Five Freedoms or similar.

This approach would recognize these as discrete if interconnected values, rather than attempt to reduce them to one core value, and it would pursue them all. It would *prohibit* policies, actions and activities if they violated core principles in any area, for example, if they caused serious environmental degradation; or violated human rights; or significantly increased inequity or entailed bad animal welfare. It would *recommend* policies to the extent that they contribute positively.

Clearly, there will sometimes need to be trade-offs, but this would be done on a case-by-case approach, and trade-offs would not be acceptable in relation to core prohibitions. So, for example, violating human rights in the interests of economic development or environmental protection, would not be permitted and nor would seriously degrading the environment in the interests of economic development and so on. If any one value *were* to take priority it would need to be environmental – as, beyond a rapidly approaching point, environmental degradation will render all other values unachievable.

Interestingly, core prohibitions in relation to the environment, are in fact quite hard to establish in more than general terms. Clearly, the degradation of ecological systems, the loss of biodiversity and various kinds of pollution will all need to be included; but identifying what actually counts as 'degradation' rather than change in a constantly changing system and at what point biodiversity loss and pollution become unacceptable is more challenging than it first appears. Nevertheless, it is not impossible. And other areas are much more straightforward. Establishing what counts as the core prohibitions in relation to animal welfare would, for example, be relatively easy. In my view, they have already been identified as something very like the Five Freedoms, the Farm Animal Welfare Council's version of these being:

1 **Freedom from Hunger and Thirst:** by ready access to fresh water and a diet to maintain full health and vigour
2 **Freedom from Discomfort:** by providing an appropriate environment including shelter and a comfortable resting area
3 **Freedom from Pain, Injury or Disease:** by prevention or rapid diagnosis and treatment
4 **Freedom to Express Normal Behaviour:** by providing sufficient space, proper facilities and company of the animal's own kind
5 **Freedom from Fear and Distress:** by ensuring conditions and treatment which avoid mental suffering.

Source: Farm Animal Welfare Council (2009); see also Chapters 6 and 7 in this book

The fourth of these is especially significant. Marthe Kiley-Worthington's version of this – ensuring that the animal has the opportunity to express all the behaviour in their natural repertoire, provided this does not cause prolonged or acute suffering to others – is particularly useful, offering real insight into how to develop appropriate husbandry systems. Either version, of course, has profound implications for current food production systems. The majority of intensive systems, for example, clearly fail to provide animals these opportunities and many violate other freedoms as well.

This is in no sense meant to be a completed approach and clearly there is much to be filled in. Meanwhile, there are two key points that sound rather abstract but whose practical implications are in fact hard to overstate. The first is the means/end point. The diamond values counter this. They are firmly rooted in an acknowledgement that other living things – people, animals, the environment – have intrinsic value. Other living things and systems are ends as well as means and should not be treated merely instrumentally. Second, and following from this, the framework proposed above is of a deontological variety rather than a utilitarian one. Within utilitarianism, an approach that underpins all forms of cost–benefit analysis and the justification of deregulated market systems, trade-offs are always possible, and as witnessed in the reality of environmental collapse, structural inequity, persistent poverty and systematically poor animal welfare – all a result of trade-offs in pursuit of narrowly conceived economic goals.

Practical implications

What would all this mean in practice? Restructuring food production systems in the light of the diamond values would mean systems that were fair to people – producers and consumers across the world – and good for human health and well-being. These would be systems in which people had a say and which we could all help to shape. They would provide high-quality, affordable, accessible food for all, in a way that did not leave an impossibly degraded environment for the future and, indeed, could readily coexist and even support a huge range of wild species. Farming would work within local ecosystems rather than seek to displace them and it would be rooted in a view of ourselves as members of a vast, complex ecological community. A pragmatic sensitivity towards others in the light of our mutual interdependence and a deep respect for others in their own right is implied by the community metaphor. And this, of course, would apply to the domestic animals in these communities, who would be treated in ways that respected them as animals, instead of as the unfeeling components of machines.

All of this would be an immense challenge, of course. The evidence from so many different angles – human health, environment, animal welfare, food security – overwhelmingly suggests that this would have to mean much less (if any) meat and dairy consumption, with far fewer animals living lives of a much, much higher quality, in environments of a much, much higher quality. But it is not impossible. As Tudge and others argue, the main tenets of this have effectively been in place in varieties of traditional agriculture practised across the world for thousands of years. This is not about going back though. It is about shaking off our brief, temporary obsession with a narrowly conceived set of economic values and moving towards a saner future. Farming systems based on respect for other forms of life – including the animals within them – should be pursued because this is the right ethics, and also because we need to think like this if we are to continue our tenancy on this, one planet. High standards of animal welfare are an abso-lutely necessary part of this. They are, or should be, a core aim of sustain-able development, which itself should be an overriding priority, given our current multiple crises. The mindset that acknowledges the importance of animal welfare as well as the other 'diamond' values is part of the solution in all these cases. And if you ain't part of the solution...

Optimistic postscript

'What in cosmos happened to you, then? Your career looked so promising. And now...'

The anthropologist shrugged.

'It was so unexpected. So utterly unlikely. But they did it. They turned it around. The Earth humans took control. Against all the odds they started to act rationally in relation to the information about the problems their food production systems were causing; and they began to respond compassionately and wisely – in their own limited ways – in relation to their impacts on other beings. Before I could say "galactic worm-hole", most humans were eating primarily vegetarian food, of a very high quality. Their health gains were massive, of course, and their media was temporarily dominated by meal-makeovers with vegetarian chefs as the new screen idols – completely displacing footballers, and even that inane vehicle worshipper, Clarkson whatshisname. Malnutrition was practically abolished from the planet and, at the same time, because of the farming methods they adopted, biodiversity increased and the integrity of some extremely compromised ecological systems was largely restored. And of course, because of the reduction in carbon emissions and the increase in natural biomass, this all had the effect of reducing the extent of climate change and, in turn, conflict over other resources, like water.'

'And what happened to "meat"?'

'It was still eaten, but only occasionally. It came to be considered a real luxury, to be savoured on special occasions. And the animals concerned were kept in immeasurably better conditions. So called "intensive farms" – more like factories than farms – became a thing of the past, as did animals kept in painful, cramped conditions, performing stereotypical behaviours and transported long distances in appalling conditions while still alive to be killed in substandard abattoirs. All gone. A much smaller number of animals living extensive, high quality lives and slaughtered locally and humanely were all that remained. Which made human agriculture vastly more efficient of course. Hence the ability to feed their population and coexist with other species. Wonderful from an earth perspective. But it completely finished me as a researcher. Examples of sane, wise, rational, compassionate thinking, albeit rare on Earth, are commonplace in the rest of the cosmos. I never got another grant after that...'

'So what are you doing now? You don't look as distressed as you might.'

'Well, oddly enough, it's all worked out rather positively. I realized academia could be a bit of a, well, to use an Earth expression, "rat-race". All that pressure to publish. And I rather took to some of the Earth conclusions about quality of life being about quality of time. I'm working half-time now, on my own book. It's loosely based on my previous research, but it's written for a wider audience. Travel. Pictures. Anecdotes. Even a few Earth recipes...'

Notes

1   See, for example, key works by these authors: Singer, P. (1975) *Animal Liberation*, Pimlico, London; Midgley, M. (1983) *Animals and Why They Matter: A Journey Around the Species Barrier*, University of Georgia Press, Athens, GA.
2   Roland Bonney has argued that an alternative and better approach is via the 'Three E's' – environment, economy and ethics, with good animal welfare included, alongside decent treatment of people, under 'ethics'. His argument is that this makes it more likely that animal welfare will be included as, once 'ethics' is admitted as a valid component of sustainable development, it is very hard to argue that animal welfare is not a relevant aspect of ethics. The 'diamond' approach in his view still leaves animal welfare vulnerable – it is easy enough to lop off or ignore the extra arm.
3   See, for example, Regan, T. (2004) *Empty Cages: Facing the Challenge of Animal Rights*, Rowman and Littlefield, Lanham, MA, and www.tomregan-animalrights. com/home.html.

References

Anderson, R. (2007) 'Mid-course correction', *Resurgence*, no 242

Bonney, R. (2008) 'Ethics in action: Farming, the environment and animal welfare', presentation at 'Reconnections', Forum for the Future, Yewfield, Cumbria

Bonney, R. (2009) 'How do livestock producers balance the demands for food production and environmental responsibility with animal welfare?' presentation at 'Animal welfare, sustainable farming and food security', AWSELVA conference at the Food Animal Initiative, Wytham, Oxford, UK

Eshel, G. and Martin, P. (2006) 'Diet, energy and global warming', *Earth Interactions*, vol 10

Farm Animal Welfare Council (2009) 'Five freedoms', available at www.fawc.org.uk/ freedoms.htm, last accessed 3 June 2009

Global Humanitarian Forum (2009) 'The anatomy of a silent crisis', available at www. ghf-geneva.org/index.cfm?uNewsID=157, last accessed 9 June 2009

Kiley-Worthington, M. (1993) *Eco-Agriculture: Food First Farming – Theory and Practice*, Souvenir Press, London

McMichael, A. J., Powles, J. W., Butler, C. D. and Uauy, R. (2007) 'Food, livestock production, energy, climate change, and health', *Lancet*, vol 370, pp1253–1263

Monbiot, G. (2008) 'This stock collapse is petty when compared to the nature crunch', *The Guardian*, 14 October, available at www.guardian.co.uk/commentisfree/2008/ oct/14/climatechange-marketturmoil, last accessed 21 April 2009

Rawles, K. (2006) 'Sustainable development and animal welfare: The neglected dimension', in Turner, J. and D'Silva, J. (eds) *Animals, Ethics and Trade*, Earthscan, London, pp208–216

Rawles, K. (2008) 'Environmental ethics and animal welfare: Reforging a necessary alliance', in Dawkins, M. S. and Bonney, R. (eds) *The Future of Animal Farming: Renewing the Ancient Contract*, Blackwell Publishing, Malden, Oxford, Victoria

Regan, T. (1985) 'The case for animal rights', in Singer, P. (ed.) *In Defence of Animals*, Blackwell Publishers, Oxford

Stehfest, E., Bouwman, L., van Vuuren, D. P., den Elzen, M. G. J., Eickhout, B. and Kabat, P. (2009) 'Climate benefits of changing diet', *Climatic Change*, vol 95, nos 1–2, pp83–102

Stern, N. (2006) *The Economics of Climate Change: The Stern Review*, Cambridge University Press, Cambridge

Stern, N. (2009) *A Blueprint for a Safer Planet: How To Manage Climate Change and Create a New Era of Progress and Prosperity*, The Bodley Head, London

United Nations Environment Programme (2007) *Global Environment Outlook GEO4 Environment for Development*, Progress Press Ltd, Malta, available at www.unep.org/geo/geo4/media/, last accessed 21 April 2009

Wilkinson, R. and Pickett, K. (2009) *The Spirit Level: Why More Equal Societies Almost Always Do Better*, Allen Lane, London

14

Religion, Culture and Diet

Martin Palmer

As the world begins seriously to take stock of the true cost of our food consumption – especially meat – the need to find powerful forces within ourselves and our cultures to help us change becomes more urgent. One of the most powerful but often most neglected is the power of religious tradition. In every faith there are millennia old traditions of simplicity, fasting and appropriate foods. Long dismissed as simply old-fashioned, superstitious or irrelevant, they are being rediscovered by the faiths and recognized by environmentalists as profoundly wise. For within these traditions lie the experience and wisdom of generations of humanity and, as far as the faiths are concerned, the wisdom of the Divine as to how we should live as part of this complex, beautiful world of creation. This wisdom relates not just to good ecology, or even good spirituality. It is also the wisdom of how to live and eat well and keep well.

In the early 1990s research undertaken by the Greek medical authorities into the reasons for heart attacks cast light on an unexpected insight. They discovered that heart attacks happened far less frequently to Greek men and women over 50 years old than to similarly aged people living in other European countries (Chliaoutaki, 2002; Sarri et al, 2003, 2004). After investigating and discarding various possible reasons for this they arrived at the conclusion that this was to do with the fasting laws of Greek Orthodox Christianity. It was shown that observing these laws and traditions ensured a far greater possibility of a long and healthy life than those who did not follow the laws and traditions.

The tradition in Orthodoxy was to fast one day a week every week; twice a week in certain periods and then three days a week during special seasons such as Lent and Advent. Fasting involved no meat; no oil; no cheese or milk products; and no wine. This means that for example during Lent – preceding Easter – the diet is effectively vegan with no oil except on Sundays and no wine.

In 2007, similar medical research into monks on Mount Athos in Greece found that these men were far less likely to suffer heart attacks or certain kinds of cancer than other European men and concluded that this was largely because of the religious fasting laws.

Fasting, abstinence and the following of specific diets is an important feature of many faiths and could be one of the most powerful sources possible for redirecting lifestyles. The revival of interest in religion and in the religious life worldwide is of such a scale as to bring new energy to old traditions, which have now been found to be of extraordinary relevance to a world that thought they were just old superstitions or practices. Not only have these practices been found to be beneficial medically – see below for further specific examples – but they are being built into the long-term plans that almost every major religious tradition is now developing worldwide to respond to the environmental crises that press upon us.

In 2009 at a joint event at Windsor Castle organized by the UN and the Alliance of Religions and Conservation (ARC), lifestyle changes were one of the three main foci for the commitments of the faiths to respond to environmental challenges. For example, Judaism, Islam, Hinduism and Sikhism each decided to start faith-based labelling systems for the faithful to promote organic, free range, more environmentally friendly production – e.g. Forest Stewardship Council (FSC) products; Marine Stewardship Council (MSC) fish supplies; and energy efficient products, while every major faith committed to purchasing ethically.

The reason this was possible and will be effective is that such traditions of conscious decision making about what to eat and how that food should be considered lies deep in every faith tradition and far back in time. In the case of Judaism and Islam this was building upon traditions of prohibition – kosher and halal – but with a twist. Kosher and halal are essentially about what not to eat. The new programmes are about what you should eat – organic, free range, fair trade etc. In the case of Hinduism and Sikhism the plans take basic notions of compassion and respect and highlight new areas of concern for those faith traditions.

For some faiths who attended Windsor and launched their long-term plans, there were no such ancient traditions of kosher, halal or Hindu vegetarianism. Instead these faiths – Christianity, Daoism and Buddhism – sought to explore monastic traditions of simple living that embody many insights about healthy and thoughtful eating – as has been highlighted by the Christian orthodox examples above.

There is therefore a vast and powerful series of forces, models and lifestyles to be explored within each faith and to do this we need to understand how these work and what their cultural roots and significance are.

There are many stories told within the myths and legends (myths here meaning truth-bearing) from diverse traditions that emphasize the need to have a diet directed by spiritual concerns. For example this is captured in the shift in diet recorded in the bible with regards to Adam and Eve and then a few generations later, Noah and his family. In the *Genesis* story of God creating a man and woman on the last day of Creation (*Genesis*, Chapter 1:29) the man and woman are told that they can only eat vegetables and fruit. This is seen, in biblical terms, as being the natural and right food for human beings, prior to the coming of sin.

However, after the disaster of the moral decline of human beings that later results in the sending of the flood by God to purge and purify the world, God allows human beings to eat meat for the first time, albeit with the beginnings of the strict kosher laws that were to come to define Judaism later (*Genesis*, Chapter 9: 2–3). The traditional interpretation of this new dietary permission is that God recognized that humanity was no longer capable of living the life he had hoped they would, and that he also recognized the fallen nature of humanity, with meat eating as a concession to this imperfect life that now existed. A corollary of this is that in Paradise, we will return to being vegetarians (if indeed we eat at all). Based on this tradition, Hazon, the world's largest Jewish social and environmental network, works with thousands of synagogues and Jewish organizations on community supported farming projects. Underlying this is a discussion about whether we now need to return to that primal diet; whether in other words we can now become responsible again in our relationships with God and with the rest of creation. One of the Jewish community's commitments is to reduce communal meat consumption by 50 per cent by 2015 (Jewish Seven Year Plan, 2009).

We thus find in one of the most influential stories in the world – a story that has shaped Judaism, Christianity, Islam and the Baha'i faiths representing some 50 per cent of all human beings and which gives many people a sense of their origin and meaning – that the issue of appropriate food, diet and abstinence is a central theme. It is a powerful indicator of how diet has always been a key issue for faiths and cultures.

Earliest references to dietary laws

The earliest detailed references we have to dietary and fasting laws occur in the Hebrew bible, especially in the Torah (the Five Books of Moses) dating from before 1000 BC, as well as in the Laws of Manu from Hinduism also known as Mānava-Dharmaśāstra, which were recorded in around 200 BC but are traced back to oral traditions before 1000 BC. From Greek and Latin sources including Porphyry of Tyre's third century book *De Abstinentia* we learn of reports about the dietary laws of Egyptian priests at that time. He quoted the philosopher Xenocrates asserting that three of his laws still remain in Eleusis, which are these: 'Honour your parents; Sacrifice to the Gods from the fruits of the earth; Injure not animals.'

The laws from the Torah and from the Law of Manu probably both date from the middle of the second Millennium BC in their earliest forms. The Jewish laws are still observed to this day by Orthodox Jews, while many of the contemporary dietary traditions and laws of Hinduism, Jainism and Buddhism have their roots in the Laws of Manu.

From the Torah, and in particular the Book of Leviticus, comes the prohibition on pork, camel, the hare and rock-badger as well as less likely food stuff such as lizards, weasels, mice and the chameleon. There is considerable debate as to why these laws came into being and what they mean. The Jewish Encyclopaedia list four main reasons:

- Hygiene: meat rots quickly in hot climates and pork in particular rots very swiftly and eating rotten meat can cause considerable illness and even death.
- Ethnic identity: the Jews set themselves apart from others by these practices and made it easier to retain their identify amongst stronger nations.
- Symbolic: the animals named are each linked to a vice so that avoidance of eating them means you avoid the vice.
- These were totem animals for neighbouring tribes and thus were taboo.

Whatever the reason, the health benefits of the Jewish diet are now a consideration, though their significance does not appear to be as strong as the fasting traditions of other faiths.

The Hindu laws in the Laws of Manu, especially in Chapter 5, are as detailed as those in the Torah of Judaism. The classic statement on why adherence to a strict diet is essential is spelt out in the opening verses:

> *The sages, having heard the duties of a Snataka thus declared, spoke to the great-souled Bhrigu, who sprang from fire: 'How can Death have power over Brahmanas who know the sacred science, the Vedas and who fulfil their duties as they have been explained by Thee O Lord?' Righteous Bhrigu, the son of Manu answered the sages: 'Hear in punishment of what faults Death seeks to shorten the lives of Brahmanas. Through neglect of the Veda-study, through deviation from the rule of conduct, through remissness in the fulfilment of duties,* and through faults committed by eating forbidden food, *Death becomes eager to shorten the lives of Brahmanas.* [author's emphasis] (Manu, n.d.)

It is clear from this early text that diet was viewed as on a par with spiritual exercises and wisdom and that it can cause the same spiritual and actual death as ignorance or wilful disobedience of the will of the gods.

The Daoist dietary laws

The long list of forbidden animals, birds and fish as well as such root plants as garlic and onions in the Laws of Manu is very close to elements of Judaism and also bears a remarkable similarity to the Daoist dietary laws. These date from around the 4th to 6th centuries AD and are concerned with the maintenance of the *qi* – the life energy that animates the body. In Daoist thought (which itself is based upon philosophical notions of the body and soul dating from the 6th to 3rd centuries BC) when we are born we have within our bodies all the *qi* we can ever have. The moment we start to breathe out we begin to lose this *qi*. Hence the dietary laws prohibit any foods that cause wind or indigestion as this means the wasteful loss of *qi* and thus an earlier death. This, coupled with almost exclusively vegetarian food, has led to many Daoists living long lives of remarkable healthiness, much commented upon by writers down the centuries. However, to the best of my knowledge there have been no scientific studies to compare with the work on Orthodox Christian monks and Greek Orthodox lay people.

When the Daoists launched their long-term plans and commitments in 2009 they drew upon this notion of *qi* and also on the Daoist philosophy of yin and yang. The relevance of this to contemporary environmental and welfare issues was powerfully captured by Olav Kjorven, Assistant Secretary General of UNDP, when he addressed the heads of all the Daoist temples in China in 2008 as they prepared their long-term commitment to protect the living planet:

> *Over the past couple of days I've had a chance to learn a bit from your tradition although I have to humbly say that I'm only scratching the surface. I have visited the White Cloud Temple in Beijing and yesterday I went to Mao Shan Temple here in this area. And I have learned that in your tradition as well, our duty as humanity is to restore balance, and to care for and respect nature – and that human wellbeing fundamentally depends on maintaining this balance.*
>
> *Let me quote from one of your holy books: Chapter 42 of the Dao De Jing.*
>
> *'The Tao gives birth to the origin;*
>
> *The origin gives birth to the two – yin and yang.*
>
> *The two gives birth to the three – heaven, earth and humanity.*
>
> *The three gives birth to all creation, all of nature.'*

This is very important teaching for our world today. I'm sure that many of you have heard of the common challenge we are facing around the world: that of climate change.

Because of the way we utilize energy and organize our economies around the world we are in danger of severely disrupting the balance of the climate, which conditions everything around us.

But what is climate change in its most simplistic scientific sense? It is all about the balance of carbon – a very important component in our natural system which makes life possible as we know it. It is about the balance of the carbon that exists in the air, in the clouds, in the atmosphere that surrounds us on the one hand and the carbon in the earth on the other, including in living things. And what we have been busily doing as humanity – particularly in the rich countries, but also increasingly in other countries around the world including China – is to disrupt that balance and move a lot of carbon out of the earth and into the clouds. This is familiar to you, I think because if this is anything at all it is the disruption of the balance between the yin and the yang.

And I think that your tradition as Daoists in China – your expression of the yin and the yang and how it relates to our existence as human beings – expresses better than any other religious tradition that I know of the challenge that we are facing when it comes to environmental degradation and climate change.

Given the role of meat production and animal by-products the yin/yang imagery speaks not just to the general issues but also to our lifestyle choices.

What becomes clear already from the three cases cited above is that dietary laws and fasting are intimately interwoven with spiritual practices. Echoes of this can still be discerned in Europe today when the Christian fasting period of Lent – the seven weeks before Easter – are still used by many who often have little or no formal contact with organized religion, as an excuse that society validates, to commence a diet. The strength of communal acceptance of a time of restraint, fasting or dieting still has a role and place even in countries where the observance of religion has waned.

The fast of Ramadan in Islam is perhaps the best known example of a fasting tradition that has maintained its power to today. The 30 days of the month of Ramadan are devoted to fasting from sunrise to sunset. This combined with the total ban on alcohol and the Halal rules on what may be eaten and what may not has created a diet and way of life that is considered very healthy. The scientific fascination with the Islamic dietary laws and especially the Ramadan fast has led to many studies of its nutritional and medical significance.

Islamic dietary laws

Fundamentally, diet is seen to be a key element of Islamic well-being. Interestingly, traditionally, the first thing that an apprentice physician would learn was how to cook – both for himself and others.

Islamic laws – Unani – being the laws associated with healthy living – provide two major openings into the lifestyles of the Muslim world that need to be more properly considered than has been the case thus far by animal welfare groups and environmental organizations:

- Because of core beliefs, Unani very strongly supports the move towards better understanding and care for food nurturing and growing because these are seen as both a gift from God but also as our responsibility because we have been appointed as khalifas – vice-regents upon Earth – and have the moral and spiritual responsibility to ensure the world functions well at all levels. A problem has been that Islam has never been properly asked to take part and indeed misunderstanding of halal has often created barriers of mistrust.
- Kindness to the food you eat: there is a Hadith (account of the words of Muhammad) that suggests that the family or community wanting to eat a chicken must let it live with them, to be taken care of, for three days. Re-emphasizing such a ruling, again, would be excellent for the free-range movement.

In Islam itself, fasting and the dietary laws are seen as essentially spiritual. Indeed, the institution of fasting in Islam comes second only to that of prayer. In the Qur'an, those who fast are called *saih* meaning spiritual travellers – see Surah 9: 112 and Surah 66: 5. Fasting is seen as preparing oneself for the spiritual struggle to remain faithful, in Islam – submission – to the Will of God. In one of the Hadiths (sayings of Muhammad) he said, 'Fasting is a shield.'

Once again the question of the ban on certain foods, especially pork, raises questions of whether we have here a sanctification of sensible hygiene laws. Islam is quite clear that it has inherited its basic dietary laws from those traditions that came before it – Judaism and Christianity in particular. The Qur'an makes this quite clear in Surah 2: 183: 'O you who believe, fasting is prescribed for you as it was for those before you, so that you may guard against evil.'

The evil is overindulgence and lack of restraint and self-possession. These are exactly the tools that we need, the entry points into cultures that we seek if we are to be serious about lifestyle choices.

One could ask: how does a culture prescribe good dietary and ecological practices? The best way is to build them into annual rituals and rites sanctioned by faith and maintained by social pressure. The question that is open to so many answers is which came first. Did religion always equate food, restraint and taboos with spirituality and by accident discovered the health benefits that we now also see as being environmental benefits; or did religion encode discoveries about a healthy diet and appropriate foods into a set of rituals that were designed as much for physical well-being as for spiritual development?

Either way, it is clear that modern scientific research is finding in these ancient traditions models for healthy sustainable living, which fascinate the scientific community.

Perhaps it is time that the environmental and animal welfare organizations paid academic attention to the ecological implications of such diets, cultures and faiths, as the traditions outlined above offer unique structural and psychological entry points for engagement with people and their lifestyles.

There is no other form of social pressure that can so well enable, enforce or assist the following of a specific diet and fasting regime. In studying the health traditions of the faiths, this significant contribution to generations of well-being should be given a special place and consideration.

References

Chliaoutaki, M. (2002) 'Greek Christian orthodox ecclesiastical lifestyle: Could it become a pattern of health-related behavior?', *Preventative Medicine*, vol 34, pp428–435

Jewish Seven Year Plan for Climate Change (2009) http://jewishclimatecampaign.org/aboutThisCampaign.php, last accessed 25 May 2010

Manu (n.d.) 'The Laws of Manu', translated by G. Buhler (1886), *Sacred Books of the East*, vol XXV, Oxford University Press, Oxford, pp169–170

Sarri, K. O., Tzanakis, N. E., Linardakis, M. K., Mamalakis, G. D. and Kafatos, A. G. (2003) 'Effects of Greek Orthodox Christian church fasting on serum lipids and obesity', *BMC Public Health*, vol 3, p16

Sarri, K. O., Linardakis, M. K., Bervanaki, F. N., Tzanakis, N. E. and Kafatos, A. G. (2004) 'Greek Orthodox fasting rituals: A hidden characteristic of the Mediterranean diet of Crete', *British Journal of Nutrition*, vol 92, pp277–284

Part 5

Devising Farming and Food
Policies for a Sustainable Future

15

Policy Strategies for a Sustainable Food System: Options for Protecting the Climate

Stefan Wirsenius and Fredrik Hedenus

In this chapter we argue that in order to substantially reduce greenhouse gas (GHG) emissions from food production and to preserve natural and agricultural biodiversity, policies that separately address the demand and the supply sides of the food system will be required. Taxes on animal food, and other policies that shift consumption patterns towards less GHG intensive and land-demanding food, will be crucial for reducing agricultural GHG emissions as well as for mitigating biodiversity losses related to the expansion of agriculture into natural ecosystems. Demand-moderating policies are vital because of the overall low potential for reducing agricultural GHG emissions by technological means, and because of the inherently large land requirements of ruminant meat (beef and lamb) production. However, demand-side policies alone are far from enough. Comprehensive supply-side policies will also be required, especially for containing agricultural land expansion in order to protect biodiversity in tropical regions. Supply-side policies, such as direct subsidies, will also be fundamental for preserving agricultural-related biodiversity in Europe and other regions holding biodiversity-rich permanent pastures. The latter holds for Europe even if no policies that moderate the demand for ruminant meat are put in place, since the low-intensive land use characteristic of these areas in either case is not economically viable in the long run. Furthermore, the biodiversity-rich areas represent a minor share of the total agricultural land in Europe. Therefore, the goal of preserving agricultural biodiversity in Europe should not be taken as a counter-argument against reducing global ruminant meat production by the implementation of demand-moderating policies.

Major sustainability issues of food and agriculture systems: Climate and biodiversity

This chapter deals with the arguably two most severe effects on the Earth systems caused by food production – climatic change and loss of natural ecosystems and their biodiversity. In this section, we briefly describe some of the basics related to these issues.

The role of food and agriculture systems in climatic change has increasingly been put on the political and scientific agenda. Even though carbon dioxide from energy and transport systems is the largest contributor to climate change, carbon dioxide from agricultural land use change and methane and nitrous oxide from agriculture represent roughly 18 per cent of global GHG emissions (Houghton, 1999; Steinfeld et al, 2006; IPCC, 2007).

At the global level, GHG emissions from food and agriculture are dominated by soil/vegetation carbon losses from conversion of forests and other land to agricultural land, nitrous oxide from nitrogen-fertilized agricultural soils and methane from the feed digestion in ruminants ('enteric fermentation'), see Figure 15.1. Emissions from energy use (on farms) and fertilizer production make up roughly 8 per cent of global agricultural emissions and are small compared to those of methane and nitrous oxide. In the European

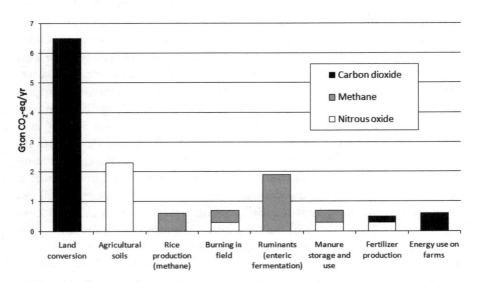

Note: All numbers are subject to considerable uncertainty, especially the carbon dioxide emissions from land conversion to agriculture.

Sources: Compiled from Koungshaug (1998); Houghton (1999); Lal (2004); Steinfeld et al (2006); IPCC (2007)

Figure 15.1 Global greenhouse gas emissions from agriculture

Union (EU), soil and vegetation carbon losses from agricultural land are negligible (or possibly even negative) and methane and nitrous oxide by far make up the great majority of agricultural GHG emissions.

Avoiding dangerous climate change involves reducing the global greenhouse gas emissions by at least 50 per cent by 2050, or even more if the climate turns out to be more sensitive to increased levels of greenhouse gases. To make this deep cut in emissions, a number of studies have shown that costs would be considerably less if not only carbon dioxide emissions were reduced but also those of other greenhouse gases, particularly methane and nitrous oxide (Reilly et al, 1999; Manne and Richels 2001; Weyant et al, 2006). A wider inclusion of methane and nitrous oxide in climate policies would have implications particularly for agriculture, since that sector is responsible for about 60 per cent of the emissions of these gases. In their review of policies to reduce agricultural GHG emissions, Povellato et al (2007) concluded that the agricultural sector potentially offers GHG abatement at relatively low costs. But they also noted that in reality the potential efficiency gains from introducing price-based instruments in agricultural production may be offset by high transaction and monitoring costs.

Of relevance for policies aiming at achieving deep emissions cuts is that there are very large differences between the GHG emissions from different kinds of food. In Figure 15.2, average greenhouse gas emissions per kcal of food for different categories are shown. The data are estimates of averages for Swedish systems, but the numbers are reasonably representative for the entire EU.

There are a number of reasons for these large differences between different food types. The main factor is the magnitude of use of land and biomass per unit of food produced – the more land and biomass that are used, the higher the GHG emissions are, since emissions are correlated with land area and turnover of crops and feedstuff. Overall, land and biomass use per unit of food is substantially higher for animal than vegetable food due to the inevitable losses in the conversion of crops and grass into animal tissue. Within the group of animal foods; land and biomass use per food unit is several times higher for beef and lamb meat than for pork and poultry, see Figure 15.3. In addition to these differences in feed required per food unit, due to the nature of the digestive system of ruminants significant amounts of methane are produced during feed digestion (of the order of 4–8 per cent of feed energy intake), which adds to the emission gap between ruminant meat and pork and poultry.

The differences in land and feed requirements are mainly because ruminants have a much lower reproduction rate (in other words, offspring produced per female parent and year) than pigs and poultry – in the EU, cows typically give birth to one calf per year, whereas sows have more

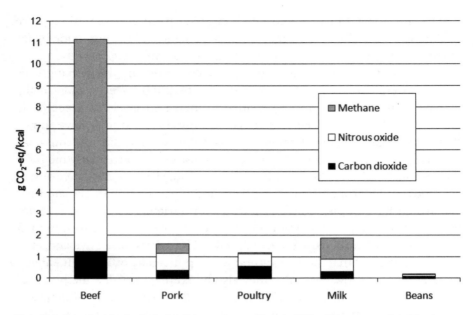

Note: Data for animal food refer to Swedish average conditions in 2005 – EU average emission levels are somewhat higher overall, but the pattern is essentially the same. Beef refers to suckler beef production.

Sources: Carlsson-Kanyama (1998); Cederberg et al (2009)

Figure 15.2 Greenhouse gas emissions per unit of food produced (in metabolizable kcal, at farm gate level) for major animal food types and beans (soil carbon balances not included)

than 20 piglets and breeding hens up to 150 chicks annually. Feed eaten by reproducing, adult females essentially represents a loss from a meat production perspective, since virtually all the feed energy is converted to body heat lost to the surroundings, and only a tiny fraction (less than 1 per cent) is converted into offspring reared for slaughter. Therefore, the lower the reproduction rate, the larger the share of total feed use (feed eaten by reproducing animals plus that by offspring reared for slaughter) that does not result in produce, in other words meat, which means that the total feed (and hence land) required for producing one unit of meat is higher.

The carbon dioxide emissions due to transportation of food often gain much attention in the media and among NGOs. However, the total energy-related carbon dioxide emissions are small as seen in Figure 15.2, and transportation is only a small fraction of these emissions. Thus, in the context of GHG mitigation in food systems, freight-related emissions are insignificant, unless aviation is used as mode of transport.

It should be noted that the numbers in Figure 15.2 do not include carbon dioxide emissions related to land use change, in other words, conversion of

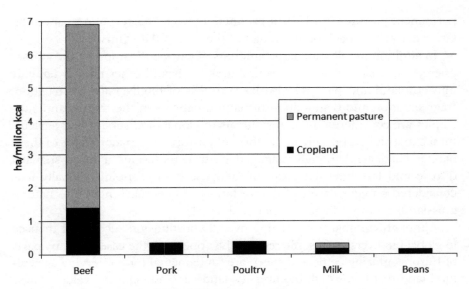

Note: Estimated averages for the EU27 in 2005. Data for beef refers to suckler beef production.

Source: Compiled from Wirsenius et al (forthcoming)

Figure 15.3 Land use per unit of food produced (in metabolizable kcal, at farm gate level) for major animal food types and beans

forests and other natural habitats to agricultural land, nor does it include the opposite process, which is sequestration of carbon in soils and vegetation. As shown in Figure 15.1, carbon emissions from land use change are substantial, accounting for about half of global agricultural GHG emissions. To what extent different food types contribute to these emissions are not very well known, due to methodological obstacles and data limitations. Yet, it is clear that ruminant production accounts for a large share, since conversion of natural habitats into pastures is a major land use change process, particularly in the tropics (Steinfeld et al, 2006).

To a significant extent, however, agricultural land use may work not only as a source but also as a sink of carbon. Studies on temperate grassland ecosystems in Europe and North America show a carbon sequestration of 200–600kg carbon per ha per year (Jones and Donelly, 2004). Recent studies of European grasslands reveal an even higher carbon sink capacity, up to possibly one ton carbon per ha per year, when well managed (Soussana et al, 2007). Although it is yet uncertain to what extent these numbers represent the true sequestration capacity over a long time (many decades), carbon sequestration in pastures and leys is likely to be an option for mitigating GHG emissions from ruminant production. For example, assuming a conservative number, 0.25 ton carbon/ha/year, of the carbon sequestration in permanent pastures in EU beef production, would correspond to about 5g CO_2e per kcal of beef, or about half of the gross GHG emissions.

In addition to releasing large amounts of carbon dioxide into the atmosphere, conversion of forests, natural grasslands and other natural habitats into cropland and pastures also leads to loss of biodiversity. Destruction, fragmentation and degradation of natural habitats in the tropics and subtropics are considered the major threats to global biodiversity, and agricultural expansion, especially in the form of pastures, is a major driver of these habitat changes. Although managed pasture represents a more spatially diffuse and less intensive form of land use than cropping, globally it is considered a major driver of deforestation and encroachment into natural grasslands and woodlands (Asner et al, 2004).

Although ruminant production overall constitutes a substantial menace to global biodiversity, in some regions the opposite is the case. In many parts of Europe, grazing cattle and sheep are fundamental for conserving agricultural land that holds high biodiversity, landscape and cultural values. These 'high nature value farmland' areas comprise hot spots of grazing-dependent biodiversity and are usually characterized by extensive farming practices. The majority of the areas consists of semi-natural grasslands, and is estimated to comprise 15–25 per cent of EU agricultural land area (EEA, 2004).

As agricultural expansion in the tropics contributes to large GHG emissions and biodiversity loss, a relevant question is whether these negative effects could be attributed only to the food produced on recently exploited land (most often ruminant meat), or all food produced in the entire region or country, or to all food produced in the world. We argue that to a significant extent they should be attributed to food products with high land requirements regardless of location of production. As demand for land increases anywhere in the world, this translates into increased global food prices, which leads to three things: food is grown more intensively, demand for food is somewhat reduced and new land is converted into agricultural land. Which of these knock-on effects is most prominent varies between regions, but to some extent increased demand for land in Europe leads to increased pressure on tropical forests, as has been illustrated by Searchinger et al (2008) in the case of increased demand for US cropland for biofuels. This phenomenon will also prevail in the long run as i) food and feed markets are largely global, with relatively low transportation costs for several major commodities (cereals, oil crops, meat), and ii) there are strong supply-side constraints due to finite land area and bio-physical limitations in yields.

Scarce agricultural land resources become an even larger challenge as the world calls for large scale climate mitigation. If stringent policies aimed at achieving deep cuts in carbon dioxide emissions from the energy system are implemented, by substantially increasing the cost of emitting carbon dioxide through, for example, taxes or emissions cap and trade schemes, demand for biomass for energy purposes is likely to increase considerably. As a price on carbon dioxide makes fossil fuels more costly than bio-energy, the demand for biomass and thereby agricultural land will increase. Estimates indicate that a stringent climate policy may increase grain prices by 50–200 per cent by 2050 (Johansson and Azar, 2007; Wise et al, 2009). The magnitude of the food price increase as well as the technical potential of bio-energy is dependent on the area efficiency in the agricultural as well as the bio-energy system.

It can be concluded that increasing the land use efficiency of food production across the board, not only in tropical regions, is crucial for preserving biodiversity in the tropics and for mitigating land use change-related GHG emissions. In the longer run, increased land efficiency in food production is also central for allowing bio-energy to expand and thereby reducing carbon dioxide emissions from the energy system in a more cost-effective way.

Options for mitigating the climate impact from food systems

Overall, options for reducing GHG emissions from food production may be divided into either technological (improved technology and agricultural practices), or structural (substitution in production mix towards less emission-intensive food types). In this section, we make some approximate assessments of each of these potentials with a focus on the EU.

For the technological mitigation options, we make a rough estimate of the potential of major options in EU food production systems by the year 2020. We consider only specific changes in technology and practices and do not take into account general agronomic efficiency improvements that may take place over time. Overall, assumptions on the diffusion of improved practices and technologies are on the optimistic side, assuming that 25–50 per cent of all farms undertake the measures. For methane, we consider altered feed ration to reduce emissions form enteric fermentation (Iqbal et al, 2008; Weidema et al, 2008) and improved manure management (DeAngelo et al, 2006). For nitrous oxide, we take into account increased efficiency in fertilizer use (DeAngelo et al, 2006), nitrification inhibitors to reduce nitrous oxide emissions from fertilizers and from manure on grazing land (Snyder et al, 2009) and pH reduction in manure to reduce ammonium emissions, which reduces indirect nitrous oxide emissions (Weidema et al, 2008). Finally, 10 per cent of the transport energy is assumed to be renewable (here, for simplicity, assumed to be climate neutral), in accordance with the EU Renewable Energy Directive (Official Journal of the European Union, 2009). It should be noted that some of these measures may be contentious –

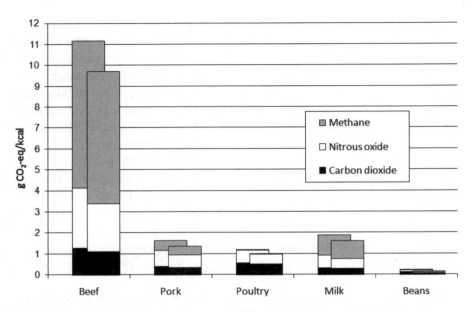

Note: Numbers in greenhouse gas emissions per unit of food produced (in metabolizable kcal, at farm gate level).

Sources: Weidema et al (2008); Committee on Climate Change (2008)

Figure 15.4 Estimated GHG mitigation potentials of improved technology and agricultural practices in EU food systems by 2020 (bars to the right, current emissions shown by bars to the left)

for instance, increasing the starch content in ruminant feed rations can be questioned from an animal welfare perspective.

If all the measures above were applied to the extent assumed possible by 2020 (25–50 per cent of operations), emissions would be reduced by roughly 15 per cent for all food types analysed here, see Figure 15.4. This magnitude of the technical potential is consistent with estimates by Weidema et al (2008) and the Committee on Climate Change (2008).

These rather limited technical potentials are related to the fact that these methane and nitrous oxide emissions occur due to intrinsic characteristics of the systems. The digestive system of ruminants inevitably involves production of methane at significant levels that cannot be drastically reduced without fundamentally manipulating the digestive process. Similarly, nitrous oxide production is an inherent part of the nitrogen cycle and a high nitrogen turnover per land area – which is required for medium to high crop yields – inevitably entails production of significant amounts of nitrous oxide.

The technical potentials estimated here do not include carbon sequestration in soils. As mentioned in previous section, the magnitude of the carbon sequestration in European grasslands may be substantial. However, the potential to actively increase the sequestration rate by specific changes in land use is largely unknown, and the long-term mitigation effect is inherently unstable since sequestered carbon at any time may be released back to the atmosphere if land use practices are changed. Furthermore, using pastures for bio-energy instead of ruminant production will in many cases offer much higher, and definite, GHG mitigation per area unit. For instance, if European grassland is used for production of biomass that substitutes for coal in power production, avoided GHG emissions would amount to some 3 ton carbon/ha/year (Wirsenius et al, forthcoming), which is several times higher than the soil carbon sequestration rate. This means that even if carbon sequestration rates in pastures could be maintained at high levels, the combined GHG mitigation effect in the beef and energy systems would still be higher if beef was produced more land-efficiently (in other words, using a smaller pasture area), assuming that bio-energy is produced on the land made available. Even though this is an efficient measure to mitigate carbon emissions, it may imply negative effects on biodiversity – clearly so in the case where unfertilized, native permanent pastures are replaced by intensive bio-energy plantations.

A crucial consequence of the limited technical potentials is that the large emission gaps between different types of food remain about the same as without mitigation, as is illustrated in Figure 15.4. This means that the potential to reduce GHG emissions by structural changes in human diets are very large, even if technical mitigation measures are implemented. For instance, if pork or poultry are substituted for beef, GHG emissions are reduced by about 80 per cent and 90 per cent, respectively. If beans containing an equal amount of protein are substituted for cattle meat, emissions are cut by 98 per cent.

The GHG mitigation potentials of changes in diets from ruminant meat towards less GHG-intensive food are even larger if we also take into account land use change-related emissions, and production of bio-energy on land made available.

As mentioned in the previous section, land use change-related carbon dioxide emissions from beef production in the tropics may be several times larger than the non-land use change emissions. Since some (about 7–8 per cent) of the beef consumed in the EU is imported from the tropics, and since there exist knock-on effects between regional beef markets, it is reasonable to assume that reduced beef consumption in the EU also leads to reduced land use change-related emissions in the tropics.

Due to the large differences in land requirements between ruminant meat and virtually all other types of food (see Figure 15.3), changing the diet away from beef reduces land use substantially, not only of permanent pastures but also that of cropland. If the land made available is used for bio-energy in order to replace fossil fuels, additional reductions in GHG emissions can be achieved. The magnitude of the mitigation effect from bio-energy varies substantially, depending on the crops used and which kind of fossil-based energy use that is replaced (see further next section).

Thus, we may conclude that in the food system changed consumption patterns hold far greater GHG mitigation potentials than does the adoption of improved technology and practices. In this respect, the food system differs fundamentally from the other major contributors to human GHG emissions, the energy and transport systems. In the energy and transport systems, new technologies, such as wind and solar power, carbon capture and storage in fossil-based power production, plug-in hybrid vehicles, biofuels, etc., have the potential to make these systems almost carbon neutral, which means that reduction of energy use by changed consumption patterns and lifestyles, although desirable, will not play the same key role as in the food system. For this reason, quite different policy options should be considered for promoting a sustainable food system as compared to sustainable energy and transport systems.

Policy strategies

Broadly speaking, environmental policy instruments can be divided into command-and-control instruments (e.g. performance standards, stipulated technology) and price-based approaches (e.g. taxes, subsidies). Within the domain of price-based instruments, there is also the choice between taxes on the emissions as such and taxes on the outputs (in other words, the produce), or inputs (energy, feedstock, etc.), that are related to, but normally not perfectly correlated with, the emissions. Taxes on emissions are generally preferable because they address directly the discrepancy between private and social cost – however, in some cases they may be less optimal and cost-effective than taxes on outputs or inputs. The latter is the case when the costs of monitoring emissions are high, the technological options for reducing emissions are limited and when the potential for reducing emissions by substitution in output or input is high (Schmutzler and Goulder, 1997).

In this section, we argue that these conditions are fulfilled for GHG emissions from food production and that, therefore, GHG-weighted consumption taxes should be imposed on animal food with the purpose of stimulating substitution towards less emission-intensive diets. This policy strategy to promote low-emission food consumption should be complemented with measures that stimulate the exploitation of the low-cost technical reduction potentials and subsidies to support grazing on biodiversity-rich permanent pastures.

Taxes on emissions have in several cases, such as carbon dioxide from fossil fuels and sulphur, proven to be cost-effective policy tools in reducing emissions. However, a prerequisite for a cost-effective application of emission taxes is that emissions can be accurately monitored. In the cases of carbon dioxide and sulphur from fossil fuels, monitoring is possible at relatively low cost. For methane and nitrous oxide emissions from farm operations, however, monitoring costs would most likely be prohibitively expensive. Emissions of methane from enteric fermentation in the digestive tract of ruminants are correlated with feed intake, but can differ considerably between individual animals, even when feed composition and other factors are similar. For instance, for cattle consuming the same feed, emissions can vary by up to a factor of two (Lassey, 2007). Therefore, to monitor methane from enteric fermentation accurately, emissions from a significant sample of animals at the farm would have to be measured regularly. Nitrous oxide emissions from agricultural soils are correlated with nitrogen fertilizer input, but variation is much larger than in the case of enteric methane, with fluctuations of several orders of magnitudes over short timescales (Bouwman and Boumans, 2002). An accurate monitoring of nitrous oxide emissions would require virtually continuous measurement for a great fraction of the fields at a farm. Obviously, for both methane and nitrous oxide the costs of such extensive emission monitoring schemes would be extremely high.

Furthermore, the effectiveness of emission taxes is lower if the technological options for reducing emissions are limited, and as was shown in previous section, for GHG emissions from food production, the technical potentials are overall minor (Figure 15.4). In contrast, the potential for reducing emissions by substitution between food products of similar characteristics (e.g. between different types of meat) is very large, as was also shown in the previous section.

We conclude, therefore, that output taxes on emission-intensive food products are likely to be a far more effective policy instrument than emission taxes for mitigating agricultural GHG emissions. An effective scheme for such output taxes could be one where the tax is imposed at the consumption level, with the tax differentiated by the GHG emission levels per unit of food. The differentiation by GHG emissions means that taxes are assumed to be weighted according to the average production emission intensities for the food categories. This means that the tax on ruminant meat should be around ten times higher than on chicken (compare data in Figure 15.2). In principle, all categories of food should be subject to a GHG-weighted consumption tax. However, it is reasonable to exempt vegetable foods, since their much lower GHG emissions per food unit mean that the administrative costs of vegetable food taxation compared to the achieved emission reduction would be much higher than for animal food.

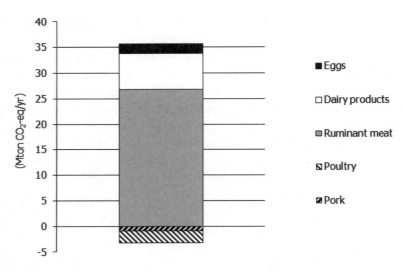

Source: Wirsenius et al (forthcoming)

Figure 15.5 Reductions in GHG emissions from animal food production in the EU for GHG-weighted consumption taxes on animal food equivalent to €60 per ton CO_2e

Assuming that such a tax scheme was introduced in the EU, significant reductions in GHG emissions would be obtained due to the tax-induced changes in food consumption. For a tax level of €60 per ton CO_2e, which is not an unlikely level of future carbon costs, assuming a stringent climate policy in the EU, total net reduction of agricultural emissions (excluding land use change-related emissions) would be about 33 million ton CO_2e – see Figure 15.5 (Wirsenius et al, forthcoming). This corresponds to about a 7 per cent reduction of current GHG emissions in EU agriculture.

GHG-weighted consumption taxes on animal food would only be considered in a world aiming at strongly mitigating climate change. In such a case, bio-energy would turn out as a profitable source of energy, since prices of fossil fuels would increase (either through a carbon tax or a cap and trade system). Therefore it is reasonable to assume that land made available due to the tax-induced changes in food consumption (see Figure 15.6) is used for bio-energy rather than remaining unused. In that case much larger emission reductions could be achieved. If the land made available was used for production of lignocellulosic crops that replace coal in electricity production, GHG emissions would be reduced by some additional 200 million ton CO_2e, or around seven times higher emissions reductions compared to those that stem from reduced animal food production only (Wirsenius et al, forthcoming). In addition to these reductions, land use change-related carbon emissions in

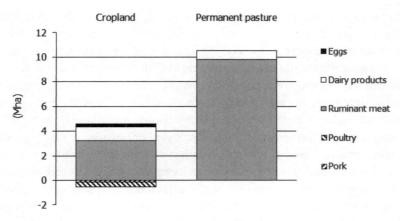

Source: Wirsenius et al (forthcoming)

Figure 15.6 Reductions in land use for animal food production in the EU for GHG-weighted consumption taxes on animal food equivalent to €60 per ton CO_2e

beef-exporting regions (e.g. Brazil) would also be reduced to some extent since the decreased beef consumption in the EU would reduce non-EU regional beef supply through global market mechanisms.

An advantage of a consumption tax instead of an emissions tax is that it would not entail substantial distortionary effects between domestic production and imports. Emission taxes imposed on production in the EU would create a cost disadvantage for EU producers in relation to producers outside the EU, which would lead to a higher import ratio in EU food supply. Although the GHG emission from EU food production would decrease due to lower production in the EU, global emissions would not decrease to the same extent, but rather move to other parts of the world (so-called 'carbon leakage'). In contrast, a tax levied at the consumption level is not likely to cause emission leakage, since the tax probably would affect EU and non-EU producers equally.

A major disadvantage of a GHG-weighted consumption tax scheme is that it does not provide continuing incentives for the producer to reduce emissions by using improved technology and practices. Therefore, a consumption tax scheme should be supplemented by policies that promote or prescribe adoption of mitigation technologies. For example, for manure storage systems, the implementation of specific mitigation options (e.g. gas proof sealing or anaerobic digesters) could be stipulated or subsidized. In this way, most of the low-cost technical mitigation potential might be captured.

Choosing policy strategies for containing biodiversity-threatening agricultural expansion, especially deforestation, in the tropics is a vast and complex topic, and it is beyond the scope of this chapter to deal with it thoroughly. Policies that moderate the demand for ruminant meat, such as the above suggested scheme of GHG-weighted consumption taxes on animal food, would mitigate the expansion pressure. However, in many areas, such as parts of the Amazon, agricultural expansion for beef and crop production is highly profitable under current circumstances (Margulis, 2003), which means that expansion will persist, albeit perhaps more slowly, even if there is a drop in global demand. At the heart of any policy for containing expansion there must, therefore, be the creation of stronger incentives to invest in already exploited land by increasing its productivity, instead of clearing and claiming new land. In many of the affected areas, lack of proper land titles is also a major factor that encourages speculative and short-sighted land use – for instance in the Brazilian Amazon only about 15 per cent of privately owned land is backed by a secure title deed (IMAZON, 2009). Introduction of policy elements such as improved land regularization and changed economic incentives also needs to be accompanied by enhanced capacity of enforcement of legislation and surveillance of illegal clearing.

The conclusion that reduced ruminant meat production is a central policy objective for mitigating GHG emissions from the food and energy systems, as well as for protecting tropical biodiversity, raises the issue of conservation of grazing-dependent biodiversity on permanent pastures in Europe and elsewhere. On the face of it, these two goals – reducing ruminant meat production and conserving pasture biodiversity – may seem to be conflicting. However, from a global perspective, the size of these hot spots of biodiversity-rich permanent pastures is small, and hence does not play a major role in world ruminant supply. This means that the biodiversity-rich pastures may be maintained also in the case of lower ruminant meat and milk production. Even in Europe, which is probably the region that holds the largest areas, they represent only a fraction of agricultural land area. Furthermore, in the case of Europe, ruminant demand at current, or even higher, levels will not prevent these areas from being abandoned, due to unfavourable economic conditions and depopulation (EEA, 2004). Therefore, to protect biodiversity-rich pastures in the EU from being abandoned, by far the most effective policy strategy is to provide specific subsidies that enhance their economic viability – rather than keeping total ruminant production at a higher level.

In the analysis in Wirsenius et al (forthcoming), the assumed GHG-weighted consumption tax leads to increased consumption of poultry and pork. This result is quite sensitive to assumptions on cross-price elasticities and available data on cross-price elasticities are sparse. Still, it is not unreasonable to assume that a GHG-weighted consumption tax would increase poultry consumption. From an animal welfare perspective this would be an unfortunate effect, since the animal welfare problems in the broiler chicken industry are well known (see Chapter 8, this book). For instance, the birds are selected to grow very quickly, which partly contributes to the relatively low GHG emissions from the production. However, the fast growth also leaves around 27 per cent of the birds lame for the last week of their lives. Thus, a more holistic approach to a climate friendly diet would be to substitute vegetables and pulses for meat, in particular beef and lamb. In that way, a diet to reduce greenhouse gases would not result in additional numbers of chickens suffering from lameness, skin damage, respiratory disease, etc. such as is prevalent in the current broiler industry.

In this chapter we have shown that the special features of the agricultural system call for an unorthodox policy mix. We suggest policies that aim at altering the diet towards fewer GHG-intensive food types. Just as policies have long aimed to increase the energy efficiency in cars by fuel taxes, it is time to also address the GHG efficiency in our diets. After all, the food system today is a larger environmental problem than the transport system. However, we have shown that these policies must also be complemented by policies that ensure a more environmental friendly production, by introducing measures to mitigate GHG emissions by technical means and subsidies to ensuring continuous grazing of biodiversity-rich permanent pastures in Europe.

References

Asner, G. P., Elmore, A. J., Olander, L. P., Martin, R. E. and Harris, T. (2004) 'Grazing systems, ecosystem responses, and global change', *Annual Review of Environment and Resources*, vol 29, no 1, pp261–299

Bouwman, A. F. and Boumans, L. J. M. (2002) 'Emissions of N_2O and NO from fertilized fields: Summary of available measurement data', *Global Biogeochemical Cycles*, vol 16, no 4, p1058

Carlsson-Kanyama, A. (1998) 'Climate change and dietary choices – how can emissions of greenhouse gases from food consumption be reduced?', *Food Policy*, vol 23, no 3/4, pp277–293

Cederberg, C., Sonesson, U., Henriksson, M., Sund, V. and Davis, J. (2009) 'Greenhouse gas emissions from production of meat, milk and eggs in Sweden 1990 and 2005', SIK Report 793, Swedish Institute for Food and Biotechnology (SIK), Gothenburg, Sweden

Committee on Climate Change (2008) 'Building a low-carbon economy – the UK's contribution to tackling climate change', available at www.theccc.org.uk, last accessed 8 July 2009

DeAngelo, B. J., de la Chesnaye, F. C., Beach, R. H., Sommer, A. and Murray, B. C. (2006) 'Methane and nitrous oxide mitigation in agriculture', *The Energy Journal*, Special Issue 3, pp89–108

EEA (2004) 'High nature value farmland: Characteristics, trends and policy challenges', EEA report No 1/2004, European Environment Agency (EEA), Copenhagen, Denmark

Houghton, R. A. (1999) 'The annual net flux of carbon to the atmosphere from changes in land use 1850–1990', *Tellus Series B-Chemical and Physical Meteorology*, vol 51, no 2, pp298–313

IMAZON (2009) 'The risks and the principles for landholding regularization in the Amazon', Instituto do Homem e Meio Ambiente da Amazônia (IMAZON), available at www.imazon.org.br, last accessed 25 May 2010

IPCC (2007) *Climate Change 2007: Mitigation. Contribution of Working Group III to the Fourth Assessment Report of the Intergovernmental Panel on Climate Change*, Cambridge University Press, New York

Iqbal, M. F., Cheng, Y.-F., Zhu, W.-Y. and Zeshan, B. (2008) 'Mitigation of ruminant methane production: Current strategies constraints and future options', *World Journal of Microbiology and Biotechnology*, vol 24, pp2747–2455

Johansson, D. J. A. and Azar, C. (2007) 'A scenario based analysis of land competition between food and bioenergy production in the US', *Climatic Change*, vol 82, pp267–291

Jones, M. B. and Donnelly, A. (2004) 'Carbon sequestration in temperate grassland ecosystems and the influence of management, climate and elevated CO_2', *New Phytologist*, vol 164, pp423–439

Koungshaug, G. (1998) 'Energy consumption and greenhouse gas emissions in fertilizer production', IFA Technical Conference, 28 September–1 October 1998, Marrakech, Morocco

Lal, R. (2004) 'Carbon emission from farm operations', *Environment International*, vol 30, no 7, pp981–990

Lassey, K. R. (2007) 'Livestock methane emission: From the individual grazing animal through national inventories to the global methane cycle', *Agricultural and Forest Meteorology*, vol 142, nos 2–4, pp120–132

Manne, A. S. and Richels, R. G. (2001) 'An alternative approach to establishing trade-offs among greenhouse gases', *Nature*, vol 410, pp675–677

Margulis, S. (2003) 'Causes of deforestation of the Brazilian Amazon', World Bank Working Paper No 22, World Bank, Washington DC, US

Official Journal of the European Union (2009), 'Directive 2009/28/EC of the European Parliament and of the Council on the promotion of the use of energy from renewable sources and amending and subsequently repealing Directives 2001/77/EC and 2003/30/EC', available at http://eur-lex.europa.eu/LexUriServ/LexUriServ.do?uri=OJ:L:2009:140:0016:0062:en:PDF, last accessed 25 May 2010

Povellato, A., Bosello, F. and Giupponi, C. (2007) 'Cost-effectiveness of greenhouse gases mitigation measures in the European agro-forestry sector: A literature survey', *Environmental Science & Policy*, vol 10, pp474–490

Reilly, J., Prinn, R., Harnisch, J., Fitzmaurice, J., Jacoby, H., Kicklighter, D., Melillo, J., Stone, P., Skolov, A. and Wang, C. (1999) 'Multi-gas assessment of the Kyoto Protocol', *Nature*, vol 401, pp549–555

Schmutzler, A. and Goulder, L. H. (1997) 'The choice between emission taxes and output taxes under imperfect monitoring', *Journal of Environmental Economics and Management*, vol 32, no 1, pp51–64

Searchinger, T., Heimlich, R., Houghton, R. A., Dong, F. X., Elobeid, A., Fabiosa, J., Tokgoz, S., Hayes, D. and Yu, T. H. (2008) 'Use of US croplands for biofuels increases greenhouse gases through emissions from land-use change', *Science*, vol 319, pp1238–1240

Snyder, C. S., Bruulsema, T. W., Jensen, T. L. and Fixen, P. E. (2009) 'Review of greenhouse gas emissions from crop productions systems and fertilizer management effects', forthcoming in *Agriculture, Ecosystems and Environment*, available online 3 June 2009 at www.sciencedirect.com

Soussana, J. F. et al (2007) 'Full accounting of the greenhouse gas (CO_2, N_2O, CH_4) budget of nine European grassland sites', *Agriculture, Ecosystems and Environment*, vol 121, nos 1–2, pp121–134

Steinfeld, H., Gerber, P., Wassenaar, T., Castel, V., Rosales, M. and de Haan, C. (2006) 'Livestock's long shadow: Environmental issues and options', FAO, Rome, Italy

Weidema, B. P., Wesnaess, M., Hermanses, J., Kristensen, T. and Halberg, N. (2008) 'Environmental improvement potentials of meat and dairy products', European Commission Joint Research Centre (JRC), Institute for Prospective Technological Studies (IPTS), Seville, available at http://ipts.jrc.ec.europa.eu/publications/pub.cfm?id=1721, last accessed 25 May 2010

Weyant, J. P., De la Chesnaye, F. C. and Blanford, G. J. (2006) 'Overview of EMF-21: Multigas mitigation and climate policy', *Energy Journal*, Special Issue 3, pp1–32

Wirsenius, S., Hedenus, F. and Mohlin, K. (forthcoming) 'Greenhouse gas taxes on animal food products: Rationale, tax scheme and climate mitigation effects', forthcoming in *Climatic Change*

Wise, M., Calvin, K., Thomson, A., Clarke, L., Bond-Lamberty, B., Sands, R., Smith, S. J., Janetos, A. and Edmonds, J. (2009) 'Implications of limiting CO_2 concentrations for land use and energy', *Science*, vol 324, p1183

16

Meat and Policy: Charting a Course through the Complexity

Tim Lang, Michelle Wu and Martin Caraher

Introduction

Meat is now becoming a 'hot' issue for the food industry, governments and consumers. Not without reason was an earlier era of US politics known as the triumph of 'pork-barrel politics'. Besides there being much money in the meat trades, meat in itself has considerable symbolic power (Rogers, 2004). To reshape meat production and consumption and to bring them in line with the Earth's capacities is a microcosm of challenges facing both the food system and the way humans live and relate with the biosphere. Because of its deep impact, meat is a test case for how and whether policy makers align the food system with sustainability goals. The evidence is strong for behaviour change, for a reorientation of production and for a refinement of supply chain management. Broadly the picture is this: first, rich societies need to eat less meat and dairy while poorer societies need to be wary about getting on to the treadmill and second, consumption is too high in developed countries and developing countries are following that lead, while all need to keep meat and dairy consumption low. As Barry Popkin and others have pointed out, there are multiple health benefits to reducing meat consumption (Popkin, 2009). And there seems to be agreement that the environmental benefits of constraining meat and dairy are high too (Sustainable Development Commission, 2009).

So why are production and consumption rising? The overall problem may be summarized as the 'meatification' of both diet and farming, an awkward word to indicate an undesirable trend. As Jeremy Rifkin noted years ago, the food system is now geared to serve cattle (Rifkin, 1992). An enormous proportion of the world's grain crop goes to animal feed rather than for direct human consumption, once again raising long-expressed concerns about inefficiency. Animal feed consumption in the EU-15 has increased by 50 per cent from 2000 with the rate accelerating sharply 2005–2007 (Eurostat, 2008). To reverse this upwards direction of meat consumption is often seen as politically explosive. This is partly an indication of meat's cultural and economic significance and partly due to understandable consumer objections to perceived restrictions on the right to choose one's diet. That's the nub of the policy challenge now facing the world's governments. There are a few exceptions; meat production is appropriate for some circumstances, climates and terrains. But the crisis we face is one of proportion; meat production is skewing policy. There is a mismatch of evidence, policy and practice, and a terrain criss-crossed by competing demands – consumer choice, culinary history, industry economic might, moral dilemmas, public health and environmental protection – which few if any politicians dare to enter.

Trying to sift what evidence matters can be tricky. It's not just the quality but the source that needs to be considered. This can enter murky waters. In Australia, for instance, Stanton and Scrinis criticized the meat component in the national research body the Commonwealth Scientific and Research Organization's (CSIRO) Total Wellbeing Diet for using research, some of which was funded by Meat and Livestock Australia; they argued that CSIRO's endorsement of a 'high-meat diet is an indication of the extent to which its scientists have taken on the role of consultants to industry in their bid to raise funding, and their willingness to deliver research findings that industry finds agreeable' (Stanton and Scrinis, 2005). The issue here is not just quality of evidence but whether funding frames the questions that get asked. Because cattle or sheep have grazed land for a few centuries, does that mean they should continue to do so? Or indeed, should have begun to?

Alas, meat is not alone in illustrating the mismatch of policy and evidence. The harmful environmental impact of current western diets is well charted but inadequately responded to. Official dietary recommendations in many developed countries still advise populations to consume at least two portions of fish a week, without reference to fish stocks being at best under stress or at worst in terminal decline (Pew Oceans Commission, 2003; Royal Commission on Environmental Pollution, 2004). And they advise us all to eat fruit and vegetables with insufficient regard to the carbon footprint of their transportation or the impact of seasonality or the coming threat of water insecurity (Garnett, 2008). With regard to meat and dairy, policy makers are not lacking in advice that to eat less would be a good thing. Lord Stern, former World Bank chief economist, whose report on the economics of climate change has been significant in adding urgency to policy makers' attention (Stern, 2006), publicly called for lower meat and dairy consumption in 2009 (Stern, 2009). Were rich economies to adopt this course – which they show little sign of doing – they might become better role models for the 21st century and be able to give some moral and political leadership (Jacobson, 2006; WCRF/AICR, 2007). While they don't, they can barely speak about, let alone address, the nutrition transition, the process whereby, as poor societies see their incomes rise, they change what they eat (Popkin and Nielsen, 2003).

Meat goes to the heart of policy makers' notions of progress. Affluence enables populations to graze more widely across the planetary food cipboard. Eating meat symbolizes economic and cultural progress. Thus meat, once only routinely available to any society's privileged, is now cheap. Unless strong cultural values such as religion or other 'rules of everyday life' such as veganism prevent or constrain it, higher meat and dairy consumption is taken politically as a symbol of rising wealth. Food that was previously exceptional, occasional and a minor part of the diet, becomes expected daily (Delgado, 2003). Meat consumption becomes a proxy for economic, social and cultural progress, and politicians are reluctant to champion meat reduction, and curb voters' aspirations. The dominant political ethos is thus to support the meat industry. In this respect, the mismatch of evidence and reality over meat is a test case for how and whether modern policy can address the complexity of food culture. Like energy-profligate housing or water-wasteful lifestyles, meat consumption and production ought to be core indicators for 21st-century change.

From the policy perspective, there is a worrying absence of dialogue between science and policy makers. Scientists are more comfortable reiterating the case for 'evidence-based policy and practice' than in listening to the evidence needs of policy makers. No wonder politicians retreat to their comfort zone, championing oft-cited market forces or consumer rights and the 'freedom to choose', as though the law of unintended consequences might not apply to those values in the real world, too. No wonder meat is seen by policy makers as an issue to leave in the 'too hard to deal with' box and to 'leave for my successor to deal with'. Yet might science help conceive of policy frameworks and food systems in which progress is defined as consuming less meat (and dairy)? Might a world be conceived where people and animals live decent lives in some kind of ecological and economic viability? As Aldo Leopold, a father of the land ethic, asked: Is meat reduction a universal goal or do animals have some use, such as in maintaining swards that retain rather than release carbon in soil? In 1925, he asked: 'Do we realise that industry, which has been our good servant, might make a poor master? Let no man expect that one lone government bureau is able – even tho it be willing – to thrash out this question alone' (Leopold, 1925).

Meat policy: A matter of framing assumptions

In this chapter, we draw upon what we and others have termed ecological public health thinking, integrating the physical or material world with the biological, social and cultural dimensions of existence (Lang et al, 2009). This territory is the interface of humans and environment often covered by the term 'sustainable development', now nominally supported by governments and in mainstream policy language (OECD, 2008). Few policy makers will overtly oppose sustainable development; dictators, oligarchs and recalcitrant neoliberal think-tanks might, of course, but not democrats or readers of long-term trends. The difficulties lie in delivery and detail (Porritt, 2005). Which macroeconomic framework (Jackson, 2009)? Which approach to health: ecological public health or productionism (Lang et al, 2009)? Where do policy reflexes lie: through technical development or social justice (de Schutter, 2009)?

Both words – 'sustainable' and 'development' – can mean diverse things, but together they denote what the 1987 *Our Common Future* report chaired by Dr Gro Harlan Brundtland, former Norwegian Prime Minister and Director General of WHO, famously defined as 'meeting the needs of the present without compromising the ability of future generations to meet their own needs' (Brundtland, 1987). Over 20 years, the sustainable development framework has gradually emerged as a reasoned way to cross-fertilize otherwise discrete policy action. Its core message is that environmental, economic and societal problems can rarely be answered in isolation; they have a tendency to knock-on to other imperatives. Single focus policies and their champions collectively store up trouble, when integrated thought and action are in fact necessary. (See Chapter 13, this book, for another view on the definition of sustainable development.)

Policy critics counter that sustainable development has little resonance with the public; it's for policy people. Let us consider momentarily the cultural aspects of meat consumption rather than its environmental or economic aspects. The association of meat eating with manliness is exhibited in many cultures; butchers are male yet the domestic task of meat preparation is more frequently female; many women eat meat, but do tend to favour the intake of their children and men over their own. Despite this central domestic role, women have barely been considered as potential change agents at a population level. The policy debates about meat tend to be pitched around climate change, economics and business.

The social aspects of meat have underperformed in meat politics, despite higher fat and meat consumption, reduced carbohydrate and fibre in the diet being features of urbanization and class relocation. Perhaps one shift that is needed in policy thinking about meat is to give more attention to this cultural dimension, to peruse meat's role in the shift from eating to survive to eating for pleasure, a move from the old to the new and modern, and the association of meat (at least for some) with affluence. In the UK, the Food Ethics Council has argued such a different position, proposing that government help consumers by building behaviour change strategies around their beliefs (MacMillan and Durrant, 2009). This position has also been proposed by a report from the Tesco-funded Sustainable Consumption Institute at the University of Manchester (Munasinghe et al, 2009).

Sidney Mintz, a food anthropologist, notes that traditional meals almost everywhere had three basic elements: a *core* food item such as rice (C), a fringe item such as a sauce (F) and a *legume* (L). This CFL pattern was even recently common in developing economies but has changed to a different recipe: *meat* (M) plus a *staple* (S) such as potatoes and two *vegetables* (V): M+S+2V (Mintz, 1996). Meat has won; it's become central rather than flavouring or exceptional. It's moved from feast-day to everyday. If the goal of a food system framed for sustainable development is to reduce meat, these social and cultural meanings and roles ought to be central and not afterthoughts.

Meat as a policy problem: Mapping the terrain

How can policy makers engage with this complexity? Many continue their long-standing support for policies to increase rather than decrease meat consumption, thus betraying their allegiance to the productionist paradigm. Yet some policy makers are beginning to support the growing policy debate about whether the world's seemingly insatiable appetite for meat can be satisfied. In 2006 the FAO published an ambitious audit in its report 'Livestock's long shadow' (Steinfeld et al, 2006). Observers pounced on it as providing evidence of the unsustainability of meat but the report also championed a policy push to reduce the CO_2 emissions from meat production without confronting high or rising meat consumption.

Food products vary widely in where their main greenhouse gas emissions are concentrated (Munasinghe et al, 2009). For cooked vegetables, it is the consumer cooking them at home that contributes most. For meat and dairy, the largest source of emissions is before the farm gate. This is why the big retailers with such a grip on milk supply chains are exploring the impact of changed feeding regimes, more efficient use of grazing (also to keep carbon in the soil rather than let it leach) and improving agricultural practices. The motive for this effort is partly self-interest – fear of being blamed later – and partly because the corporate sector recognizes that while governments and policy makers come and go and are shaped by electoral cycles, they and their shareholders have an eye on long-term market growth and share. Ironically, these champions of consumer choice are actually exhibiting 'choice-editing', a different approach to behaviour change. In choice-editing, the retailer drives change before the consumer sees the food product. The consumer's 'right to choose' is being reframed by powerbrokers in the food system.

How and where might policy makers engage with meat? We have already suggested more attention to the social and cultural aspects of meat in policy. Policy mapping is required for the core clusters of interest already vying for attention. These include: environment, health, economic, culture, national identity, ideology and philosophy, the role of science, technology and research, and the role of consumers, to all of which we now turn.

Environment

Policy makers know that meat production is associated with serious safety risks. For over two decades, UK politics was peppered with animal health-related incidents: salmonella in eggs exposed hidden food poisoning rates (Smith, 1991), BSE (mad cow disease) exposed unsavoury feeding practices (van Zwanenberg and Millstone, 2005), foot and mouth disease exposed poor farm practice (Royal Society, 2004) (and some hints of illicit trade), deaths induced by E. coli showed poor butcher hygiene standards (Penningon, 2003, Pennington, 2009). These were initially downplayed as regrettable but inevitable, but eventually recognized as more systemic. Similarly in the US, the Centers for Disease Control and Prevention estimate that 76 million cases of food-borne illness happen each year. One wonders why policy makers were content to remain in a reactive role so long. Public pressure changed that.

Policy makers are currently at the pre-action level with regard to meat's environmental impact. If at all, it is food companies who are more exercised by meat's ecological impact than governments but their thinking tends to focus on improving efficiency. Yet even relatively efficient feed converters such as poultry or pigs have environmental footprints. Policy has been more interventionist with regard to pollution such as animal effluents, not least because they can spread disease. At the Copenhagen COP-15 talks in December 2009, meat's impact on climate change featured more on the fringes than in the settlement. This policy silence might change. The UK, for instance, passed legally binding reduction targets on emissions in its Climate Change Act 2008. Such legislation brings meat and dairy into play. The evidence is there to be used (Tukker et al, 2006). Some NGOs, such as Friends of the Earth, Compassion in World Farming and WWF, have launched important public education events and campaigns but these currently raise political worries about 'nannies' dictating what people can eat.

Public health

Meat is not all bad news. Although not essential, there can be nutritional benefits from the inclusion of some meat in the human diet (Monsen, 1988; Neumann et al, 2007; Fairweather-Tait, 2007). Red meat is one of the richest sources of iron, along with minerals and vitamins such as zinc and vitamin B12 (Fairweather-Tait, 2007). This may be especially relevant for the millions of people who lack adequate food, for whom animal sources provide the most usable (or bio-available) form of many nutrients. On the counter side, there is now strong evidence of the adverse impact of high meat consumption, including higher risk of obesity and increased mortality rates due to cancers and cardiovascular disease (WCRF/AICR, 2007; Sinha et al, 2009). This has led to international guidance to cut consumption (WCRF/AICR, 2009). There is now a growing body of evidence about how to facilitate behaviour change, should policy require it.

Meat and dairy have historically been approved by mainstream health policy. 1930s nutrition thinking positively encouraged an increase in production and consumption, particularly milk for children (British Medical Association, 1939; Boyd Orr, 1943). That case is now deeply embedded; to reverse it would be a policy turn around that the public health professions have not really acknowledged. As the evidence of the negative effect of dairy fats emerged in the 1980s, the meat trade unleashed a furious reaction. Marion Nestle has documented how attempts to update US nutrition guidelines on meat and dairy were subject to intense political lobbying and mainstream policy scrutiny (Nestle, 2002). In the USA, the Johns Hopkins University's Bloomberg School of Public Health backed a 'Meatless Monday' initiative, resurrecting a wartime campaign launched at the US public and schools alike but met lobby warnings that pupils would not get enough protein, despite evidence that protein requirements were fully met.

Official nutrition advice in many countries has been to 'eat leaner meat and a bit less cheese'. In fact, the meat trade in affluent countries began to try for leaner meats in the 1980s in response to the criticism about saturated fats, but two decades later consumption remains high, and the evidence of its connection to non-communicable diseases (NCDs) remains strong (WCRF/AICR, 2007). This NCD picture has entered western policy debates about meat, but it has not dented production. Although meat's role in communicable diseases captured western policy imagination – particularly regarding BSE – its impact is greater in the developing world (Nierenberg, 2005; McMichael et al, 2007). Meat is also associated with communicable diseases such as salmonella, campylobacter and E. coli. In the 2000s, swine flu emerged as a public health concern, unleashing a huge global collaboration, and renewed need to track zoonotic diseases (European Academies Science

Advisory Council, 2008). Bio-security now stands alongside risk assessment and management in the policy lexicon. This is welcome but perhaps too much policy attention is at the level of creating better monitoring and research. They are needed, of course, but the notion of prevention deserves higher priority.

Political economy

Policy makers claim that increased meat production and its reduced price has been one of its greatest successes since the Second World War. But this has come at some cost to taxpayers in the form of state subsidies. A recent WHO study quantified the impact of EU Common Agricultural Policy (CAP) on mortality from cardiovascular disease. Conservative estimates of mortality attributable to CAP subsidies for dairy and meat was approximately 9800 additional CVD deaths and 3000 additional stroke deaths within the EU, half of them premature (Lloyd-Williams et al, 2008). Similarly, a US study attributed 40 per cent of the recent rise in weight to lower food prices brought about by agricultural innovation (Lakdawalla and Philipson, 2002). Popkin has shown that in countries in transition a small reduction in the price of fats has huge implications across the population. This is aided by increased urbanization, which makes supply easier and introduces economies of scale (Popkin, 1998; Popkin and Gordon-Larsen, 2004). Such transitions are occurring within shorter and shorter time periods. Even here there are inequalities, as over 1 billion of the world's population can be classed as poor, relying on grain for food and local biomass for cooking. Meat introduces an additional burden in storing, cooking and preparation for such groups.

The productionist policy framework that unleashed the agri-food revolution of the second half of the 20th century hinged on generating mass production (scale) and lowering prices to consumers (Lang and Heasman, 2004). The model assumed that high cost and poor affordability and output were barriers to health and therefore that the pursuit of economic efficiency and productivity would deliver both health and public good. Half a century on, we now know that lowered costs to consumers have come with unaccounted costs; no one has paid actual money (yet) for climate change, but the expense is now being calculated. One study cited by the Pew Commission on Industrial Farm Animal Production calculated that US industrial farm animal production facilities cost US taxpayers over $38 billion in externalized costs, about $159 for each US inhabitant. The Commission argued that policy makers will almost inevitably have to curtail US consumer choice by re-internalizing externalized costs, for instance restricting the use of antibiotics that facilitate intensive feedlot systems (Pew Commission on Industrial Farm Animal Production, 2008). Reactions to the 2008 Pew Commission report showed the formidable economic leverage of the US meat trades that extends wider than animal farmers as such. A constellation of economic agents work around them including breeders, breed societies, compound and feedstuffs makers, traders, equipment manufacturers, processors, logistics companies, retailers and caterers, not to mention the pharmaceutical industry. This combined power argued against the Pew Commission's recommendations for greater controls on meat production.

Such corporate power has long been known, perhaps never more starkly exposed than by Upton Sinclair's still shocking 1906 exposé of the Chicago stockyards and processors (Sinclair, 1985 [1906]). Sinclair alleged the industry was characterized by low morals and ruthless processes, even claiming that a processing factory failed to halt production when a worker fell into machinery. Ostensibly, *The Jungle* was a novel, but a scandal erupted on publication and President Theodore Roosevelt, suspicious of Sinclair's radical agenda, ordered a secret inquiry that not only confirmed Sinclair's account but indicated that he had perhaps understated his case. As a result, US Congress passed the Food and Drug Act that up until then had been effectively blocked by industry, leading to the creation of new state institutions to deliver change. It took a 'novel' to narrow the gap between evidence, policy and institutional engagement.

The BSE crisis in the UK and Europe is perhaps a modern parallel (Phillips et al, 2000; van Zwanenberg and Millstone, 2005). When the enormity of mad cow disease dawned on the UK public, its impact helped transform policy. New laws and regulations followed (the Food Safety Act), new institutions were created (the Food Standards Agency plus committees) and a new approach was adopted to transform supply chains, with the adoption of Hazards Analysis Critical Control Point (HACCP). It is possible that the adoption of life cycle analysis (LCA) techniques might do this for meat. But, unlike HACCP, which requires a wide range of workers in the food chain to be involved, LCA requires top-down scientific expertise. The pioneering attempt to harmonize methodology by a multilateral group hosted by the British Standards Institute illustrated the complexity of doing calculations even for a 'simple' food product (involving butter) such as a croissant (British Standards Institute, 2008). A different policy avenue has been highlighted by Goodland and Anhang, World Bank environment specialists (Goodland and Anhang, 2009). They argue that advising consumers is a failed policy strategy and that more attention needs to go to highlighting how financial investment in meat is now risky, and that companies seeking long-term growth ought to invest in alternatives. This is more likely to speed up change. We must recognize that, inherently, economic reductionism in policy making will favour economic growth at all costs. Will we accept that issues with significant social, health and environmental implications are being decided primarily on the basis of profit and price?

Culture and national identity

Meat is an indicator of societal as well as individual or family status and progress. Even in cattle-based cultures such as in some southern African or Latin American states, the number of head of cattle indicates wealth. But for non-landed consumers, the aim is not owning animals but eating them. The nutrition transition analysis has shown how, as wealth rises, food previously associated with scarcity becomes available more routinely. The generosity of peasant societies, in which an animal is slaughtered for ceremonial or exceptional occasions, becomes replaced by a society for whom meat consumption is normalized and unexceptional. The policy relevance of this cultural role for meat – its meanings, its place in everyday life – cannot be underestimated. In Judaeo-Christian culture, the phrase 'killing the fatted calf' is associated with the return of the prodigal son; globally, meat was once for the unusual or the feast day. As many writers on meat have noted, meat has now lost its cultural significance.

Policy makers are highly sensitized to this cultural dimension of meat. One entry point much used to pioneer change is public food service, particularly school food. Attempts to change the quality of UK school meals, despite early support from a popular celebrity chef, Jamie Oliver (Oliver, 2006), met with passive resistance; uptake of school meals dropped. Guidelines produced by the School Food Trust, created by the government, specified that only one meat product from four categories should be served per fortnight (School Food Trust, 2009b).

Meat does not just have cultural associations with well-being and gender (especially masculinity) but also national identity. Consider the British association with 'John Bull' and beef (Rogers, 2004), or the French with camembert (Boisard, 2003). The crisis over BSE in the UK was, according to some, as much to do with national identity and the threat inherent to this as it was to do with the safety of food (Ravallion, 2004). Rogers has offered an historical account of the fondness of the British for beef, using France as a counterpoint. Similarly, Steven Mennell's magisterial exploration of food habits comes down to a comparison of British and French ways of cooking and eating, which are metonyms for the respective cultures (Mennell, 1996). Other accounts have been given of how beef became 'food of the gods', spawning its place in myth and legend, and of its role in the US pioneering spirit as settlers headed westwards with wagon trains and herds of cattle

(Rimas and Fraser, 2009). These accounts of meat and muscle highlight the way in which the benefits of meat consumption are portrayed. This is a view shared by Albritton in his exposure of the food industry and the 'meatification' of the food system both in the production of meat but also culturally in how it processes the products of the meat industry (Albritton, 2009).

This national identification with meat extends to its place in the everyday diet or meal. Not being able to afford meat in a culture where this is the norm is seen as an indicator of relative poverty. The National Anti-Poverty Strategy in Ireland uses the following indictors along with an income standard as part of its measure of 'consistent' poverty, combining a relative income measure with a composite deprivation index of eight items, three of which relate to food: having a meal with meat, fish or chicken every second day; having a roast or its equivalent once a week; not having gone without a substantial meal in last two weeks (Government of Ireland, 2002). Not having access to, or not being able to afford, fruit is not seen as an indicator of deprivation, surely an affront to national identity and sense of worth.

Australia, North America and to some extent South America are examples of cultures based on imported modes of agriculture for whose image meat has been central. The Australian food system was largely shaped by the British, who would not learn or develop the aboriginal food system based for millennia on engagement with that fragile landscape. The imposition of cattle on the Australian continent was an attempt to recreate an English/Irish/Scottish idyll. In his gastronomic history of Australia, Symons proposed that thereby Australia shifted from a hunter-gatherer to an industrial food society, skipping the agrarian model in which families plant crops around a homestead (Symons, 1998a). Yet some of that more desirable 'agrarian model' was in fact donated to Australia by other southern European migrants to Australia who brought a horticultural emphasis to their back yards and gardens (Gaynor, 2006). But dominant Australian food culture focuses on meat; the ubiquitous barbecue – the iconic 'barbie' – thus reflects not only Australia's lack of indigenous peasant culture, but also the demolition of the aboriginal identity. European settlers lived on imported and transported rations that consisted of *'ten, ten, two and a quarter'* of flour, meat, sugar, tea and salt respectively. Ten pounds of meat (4.5 kilos) seems a lot but set the benchmark for Australian food culture (Symons, 1998b).

Ideology and philosophy

The moral issues around meat have long been explored by philosophers, from the Greek ancients appealing to reduce cruelty, to Peter Singer in his 1975 *Animal Liberation* appealing to cease eating it altogether (Singer,

1975). The debate whether to abstain or refrain, to eat none or eat restrict-edly, to appeal or frame choice, has been mostly pitched at the individual level. At one level, this is a matter of ethics: moral choices about how to live, the engagement with what food ethicist Michiel Korthals has identified as 'ethical dilemmas', how to locate food in everyday life in an upstanding way, how to be overt about the ethical foundation of aspiring to live a morally good life (Korthals, 2004). Modern meat raises deep questions: not just whether to eat it or how much, but how it is produced, where and at what ecological cost? Modern meat has both deepened and widened the philosophical questions beyond the individual to the planetary, extending the Malthusian debate about whether the capacity to increase food production could outpace the capacity of human populations to increase (Malthus, 1798). Malthus' *An Essay on the Principle of Population* is arguably one of the most influential and persistent theories of the last two centuries, influencing the entire oeuvre of agri-food science whose response was to unleash industrial agriculture by incorporating Mendelian genetics on to von Liebig's and Benet Laws' application of chemistry to farming.

Both Marx (with his argument that society needs to change to obviate individual moral dilemmas) and modern science and technology (with their promise to unleash nature's potential) side-step individualized morality. Their combined pressure – revolution and democracy, genetics and industrial productivity – have enabled Malthus to be proven wrong so far. Food output has kept up with global population growth (Defra, 2009). There is ample food, as measured calorifically, to feed the world at present, if it was equitably distributed and if waste was reduced. Those are big 'ifs'. Malthus' challenge was interpreted by some as questioning whether we should try to produce more, whether the social and natural order could be pushed back, whether in fact we should accept his 'principle' as reality. This is not just a technical matter but a political and philosophical one. Should societies try to improve the human condition or is it fixed? Not so far below the surface of some environmental thinking is an acceptance and conservatism. But can championing 'living within environmental limits' (the wording of the UK Government's commitment to sustainable development in its 'Securing the future' White Paper, H.M. Government, 2005) be translated into practical policy?

No mainstream political party has yet championed radical meat reduction other perhaps than the Greens in Europe, but even they allow for choice, and they tend to espouse modes of production deemed to be 'softer' or more ethical such as organics. Meat-free Mondays and wartime rationing are important forays into this policy territory. 'Less but better quality' might be the summary of such policy positions (Singer, 2009). Within the marketplace, this becomes translated into just 'better quality', as retailers champion high quality plus high price market niches. Governments in Europe at least have not just accepted but validated this approach, with standards being set and harmonized by the EU. The individualization of market relations, however, does not quite bridge the ideological gap between broad ethical values and the continued economic reality of land use wedded to meat production. The role of NGOs here is important, acting as they do so often as scouts for public policy, testing and probing and daring.

What can policy do about these problems?

Policy responses have tended to be hesitant and low key. They amount to policy maintenance rather than redirection. Messages directed at the individual are primarily framed around responsible consumerism – choosing to buy organic or eat local – while relatively little attention is focused on the need for systemic change. Inclusion of the agricultural sector in carbon pricing, redirection of subsidies and harnessing the power of public procurement to reshape demand are all examples of policies that could help reduce meat and dairy production and consumption. Even when public outcry offered an opportunity for change – such as when in the UK the School Food Trust was set up in the wake of celebrity chef Jamie Oliver's Upton Sinclair-like exposé of the lamentable state of school food in 2006 – government has failed to take the opportunity to recommend a reduction in meat (School Food Trust, 2009a). Meat protectionism is the default policy position. It has been left to NGO initiatives to take the lead, such as WWF's One Planet Living initiative, which calls for 15–20 per cent reduction in meat and dairy consumption by 2020 for a country like the UK (WWF–UK, 2009; MacMillan and Durrant, 2009). And the UK organic movement's Food for Life Catering Mark, at gold level, suggests offsetting the higher costs of more sustainably produced meat by eating less (Food for Life Partnership, 2008).

Where might meat (and dairy) production and consumption be headed? The theoretical range of strategic options are summarized in Table 16.1.

Such moves will only develop and consolidate through the democratic process, within and outside formal governance structures. Indeed, such processes are emerging. From civil society is emerging an important quasi-formal cohort of foundation-funded inquiries, giving expression to NGOs and academic concerns. The Food Ethics Council hosted one in the UK (MacMillan and Durrant, 2009) and the Pew Commission another in the US (Pew Commission on Industrial Farm Animal Production, 2008). This suggests a growing consensus that meat should rise up the policy agenda. The issue of tactics – how to argue as well as what – has emerged in private already. The Pew Commission focused on the environmental, economic and health implications of industrial modes of animal production, reflecting a widely

Table 16.1 *The range of strategic options for meat futures*

Option	Intention	Comment
Increase production	Build meat industry and encourage consumption	This is happening but storing up future trouble
Technical development	Meet increased demand through new technology, taking a variety of forms from laboratory grown meat or intensive fish tanks to novel plant proteins	This approach requires consumer acceptance and carries risks to trust and market stability
'Freeze' at current levels	Maintain status quo	This exposes policy makers to accusations of policy drift and complacency; offers little public interest gain but is the default position for meat trades
Reduce production	Ration consumption by various means including pricing and taxation	This would raise prices but heighten unequal access, possibly increasing desirability
Reduce consumption	Stimulate change to more sustainable diets, sending signals from consumers to supply	This implies that supply chain would not respond to increase uptake
Ban or rationing	Reduce negative impacts drastically	Enforced veganism is politically unacceptable even in vegetarian cultures

held view that industrial farming is the Achilles heel of US meat culture. But do those arguments fit Europe or developing regions? Possibly not.

There is the issue of tactics: if evidence of the need to shift policy direction is so strong, is it better to promote step change or incrementalism? To wean consumers off ubiquitous meat, is it best to suggest a 'meat-free Monday' – an individual behaviour oriented, secular take on Christianity's former meat-free Fridays – or to encourage change through the re-formulization of processed foods and public service food going meat-free, or both? These questions need careful thought, experimentation, research and evaluation. Like most analysts, we favour a reorientation of policy to reduce both production and consumption. Environmentalist Jeremy Rifkin (Rifkin, 1992), ethicist Peter Singer (Singer, 1975; Singer and Mason, 2006) and consumer health campaigner Michael Jacobson (Jacobson, 2006) have each made powerful cases for what we might term a paradigm shift. Policy makers ought perhaps to consider how these can be coalesced in one set of overarching goals (see Table 16.2).

In theory, policy makers have a wide range of measures and instruments available to help them deliver policy. Their choice and use is usually shaped

Table 16.2 *Possible main goals for a meat reduction policy*

Goal 1	Reduce output and consumption of meat and dairy in developed countries
Goal 2	Halt upward trend in production and consumption in developing countries
Goal 3	Transform existing production to more ethically and sustainable modes
Goal 4	Reposition meat and dairy consumption as exceptional rather than everyday foods
Goal 5	Internalize full social, health and environmental costs into consumer prices

by circumstance and the balance of popular opinion, or what politicians dare to do. The trans-fat bans in Denmark and New York City have shown how evidence-based policies can be implemented to protect the consumer, and not just the average consumer but subgroups within the population who may consume higher levels of saturated fat. This approach is standard in public health (Rose, 1992). Where will a local authority leader dare to impose extra charges on meat – akin to London's traffic congestion charge, which cut car use by imposing a daily rate? Where will a hospital or company take the lead in cutting meat from its canteen? Options need to be tried, if real policy effectiveness is to be evaluated. Circumstances, of course, often provide natural experiments. Wars or dislocations due to health crises offer such occasions. In times of crisis, a broader range of measures tends to be politically more acceptable than in times of peace. An indicative list can be drawn, ranging from 'soft' at the top of the table to 'hard' measures towards the bottom (see Table 16.3). They also range in orientation from individual to population effect.

Some northern European countries have begun to produce relevant policy documents. In 2009 the Swedish National Food Administration and Environmental Protection Agency collaborated to produce guidance on environmentally friendly but healthy diets (National Food Administration, 2009). In 2008, the Dutch Ministry of Agriculture, Nature and Food Quality produced a policy commitment to develop a sustainable food system including reshaping consumer behaviour (Ministry of Agriculture, 2008). The message was that sustainability is about efficiency. It hinted at the case for less meat but backed off, citing the value of animal production in many developing countries. In Germany, the Council for Sustainable Development has long produced a guide *The Sustainable Shopping Basket* (German Council for Sustainable Development, 2008). This states clearly and simply 'your shopping basket should contain … less meat and fish' (p11).

In the UK, which like The Netherlands, is a big meat producer, the relevant government ministry (Defra) has been reluctant to specify meat reduction, but has acknowledged the need to reduce meat's emissions. The Sustainable Development Commission, the UK Government's advisor, formally made

Table 16.3 *The range of public policy measures available to shape meat supply and consumption*

Measure	Main sources	Implications
Advice	Tends to be state or companies	Tends to be weak and with low impact
Labelling	State or business	Puts onus on consumers. Can suffer from information overload
Education	Used to be state, but increasing presence of corporate materials	Long time to be effective; works best when coupled with other measures
Public information	Corporate. Sometimes funded by states or levies on trade	Ranges from advertising and marketing to virtual and web-based media
Endorsement and sponsorship	Corporate	Increasing use of celebrity. Some blurring of lines between media content and advertising
Welfare support	State	Tendency to use this to subsidize surplus disposal
Product/ compositional standards	Was the preserve of the state. Now used by states, supply chains (through contracts) and civil society	Rise of animal welfare and organic farm movements has had a significant effect on championing process orientations in standards setting
Licensing	Traditionally state, but now used by companies and by NGOs negotiating their own standards	Brands are licences
Subsidies	State	Deeply opposed by theoreticians (e.g. OECD) as market distorting.
Competition rules	State	Many rich societies have competition bodies that conduct inquiries and have leverage, e.g. through fines
Taxes and fiscal measures. Can include incentives (e.g., funding to develop innovative practices/products)	State	The most feared measure by corporates, as they add direct costs. Critics see them as distortions
Bans	Used to be preserve of state, but increasingly championed by corporate. Civil society organizations frequently call for them	Rise of overt corporate standards has seen 'choice-editing' being championed

Table 16.3 *The range of public policy measures available to shape meat supply and consumption (contd)*

Measure	Main sources	Implications
Rationing	Preserve of state	Tends to be used in times of war in free societies. Markets of course 'ration' by creating equilibrium between supply and demand

tougher recommendations that government champion meat reduction in late 2009 (Sustainable Development Commission, 2009). Such policy documents suggest that the notion of sustainable diet might be the terrain on which a new policy framework for meat is based. This might bring together the various initiatives by companies, governments and civil society bodies that agree that current levels of meat production and consumption cannot be sustained. Meanwhile, it has to be concluded that policy generally lags behind the evidence, and policy makers lack evidence that could help them frame policy shifts.

When he argued that future food culture should centre on the simple 'rule' of 'Eat food. Not too much. Mostly plants', journalist Michael Pollan articulated a simple, perhaps overly simple recommendation for a sustainable diet (Pollan, 2008). We too see the role of public policy in realizing sustainable food systems. However, the crisis of meat and dairy consumption is really part of a wider discourse: how we humans are exceeding our ecological niche and shaping the planet at considerable cost. An ecological public health perspective on food in general and meat (and dairy) in particular challenges us to reorient our values and identity. How can we live as citizens of a community that includes land, water, air and animals – as opposed to our prevailing identity as consumers? It is on that policy terrain that consensus needs to be built, rapidly but democratically.

References

Albritton, R. (2009) *Let Them Eat Junk: How Capitalism Creates Hunger and Obesity*, Pluto Press, London

Boisard, P. (2003) *Camembert: A National Myth*, University of California Press, Berkeley CA

Boyd Orr, S.J. (1943) *Food and the People. Target for Tomorrow No 3*, Pilot Press, London

British Medical Association (1939) 'Nutrition and the public health: Proceedings of a national conference on the wider aspects of nutrition', 27–29 April 1939, British Medical Association, London

British Standards Institute (2008) 'PAS 2050 – Assessing the life cycle greenhouse gas emissions of goods and services', British Standards Institute, London

Brundtland, G. H. (1987) *Our Common Future: Report of the World Commission on Environment and Development (WCED) chaired by Gro Harlem Brundtland*, Oxford University Press, Oxford

De Schutter, O. (2009) 'The right to food and the political economy of hunger', 26th McDougall Memorial Lecture, Opening of the 36th Session of the FAO Conference, FAO/UN Special Rapporteur on the Right to Food, Rome, available at www.srfood.org/images/stories/pdf/otherdocuments/20091118_srrtf-statement-wsfs_en.pdf, last accessed 25 May 2010

Defra (2009) 'Food security assessment', available at www.defra.gov.uk/foodfarm/food/security/assessment.htm, last accessed 27 October 2009

Delgado, C. L. (2003) 'Rising consumption of meat and milk in developing countries has created a new food revolution', *Journal of Nutrition*, vol 133, pp3907S–3910S

European Academies Science Advisory Council (2008) 'Combating the threat of zoonotic infections', EASAC/Royal Society, London

Eurostat (2008) 'Food: From farm to fork statistics', Eurostat, Brussels

Fairweather-Tait, S. J. (2007) 'Iron nutrition in the UK: Getting the balance right', *Proceedings of the Nutrition Society*, vol 63, pp519–528

Food For Life Partnership (2008) 'Food for Life Catering Mark', Bristol, Soil Association, Garden Organic, available at www.foodforlife.org.uk/resources/catering/catering-mark, last accessed 25 May 2010

Garnett, T. (2008) 'Cooking up a storm: Food, greenhouse gas emissions and our changing climate', Guildford, Food and Climate Research Network University of Surrey

Gaynor, A. (2006) *Harvest of the Suburbs: An Environmental History of Growing Food in Australian Cities*, University of Western Australia Press, Crawley

German Council For Sustainable Development (2008) *The Sustainable Shopping Basket: A Guide to Better Shopping*, 3rd edition, German Council for Sustainable Development, Berlin

Goodland, R. and Anhang, J. (2009) 'Livestock and climate change: What if the key actors in climate change are … cows, pigs and chickens?', *World Watch*, November/December, pp10–19

Government of Ireland (2002) 'Building an inclusive society. Review of the national Anti-Poverty strategy under the Programme for Prosperity and Fairness', Dublin, Department of Social and Family Affairs

H.M. Government (2005) 'Securing the future: Delivering UK sustainable development strategy', Cm 646, London, H.M. Government

Jackson, T. (2009) *Prosperity without Growth: Economics for a Finite Planet*, London, Earthscan

Jacobson, M. F. (2006) 'Six arguments for a greener diet: How a more plant-based diet could save your health and the environment', Center for Science in the Public Interest, Washington DC

Korthals, M. (2004) *Before Dinner: Philosophy and Ethics of Food*, Springer, Dordrecht

Lakdawalla, D. and Philipson, T. (2002) 'The growth of obesity and technological change: A theoretical and empirical examination', NBER Working Paper No 8946, New York, National Bureau of Economic Research

Lang, T. and Heasman, M. (2004) *Food Wars: The Global Battle for Mouths, Minds and Markets*, Earthscan, London

Lang, T., Barling, D. and Caraher, M. (2009) *Food Policy: Integrating Health, Environment and Society*, Oxford, Oxford University Press

Leopold, A. (1925) 'A plea for wilderness hunting grounds', *Outdoor Life*, no 5

Lloyd-Williams, F., O'Flaherty, M., Mwatsama, M., Birt, C., Ireland, R. and Capewell, A. (2008) 'Estimating the cardiovascular mortality burden attributable to the European Common Agricultural Policy on dietary saturated fats', *Bulletin of the World Health Organization*, vol 86, pp497–576

MacMillan, T. and Durrant, R. (2009) 'Livestock consumption and climate change: A framework for dialogue', Food Ethics Council, Brighton

Malthus, T. R. (1798) *An Essay on the Principle of Population, as it Affects the Future Improvement of Society with Remarks on the Speculations of Mr. Godwin, M. Condorcet and Other Writers*, printed for J. Johnson, London

McMichael, A. J., Powles, J. W., Butler, C. D. and Uauy, R. (2007) 'Food, livestock production, energy, climate change, and health', *The Lancet*, vol 370, pp1253–1263

Mennell, S. (1996) *All Manners of Food: Eating and Taste in England and France from the Middle Ages to the Present*, University of Illinois Press, Chicago, IL

Ministry of Agriculture (2008) 'Policy document on sustainable food: Towards sustainable production and consumption of food', Den Hag, Ministry of Agriculture, Nature and Food Quality, available at www.minlnv.nl/portal/page?_pageid=116,1640321&_dad=portal&_schema=PORTAL&p_file_id=39545, last accessed 25 May 2010

Mintz, S. (1996) *Tasting Food, Tasting Freedom: Excursions into Eating, Culture and the Past*, Beacon Press, Boston

Monsen, E. R. (1988) 'Iron nutrition and absorption: Dietary factors which impact iron bioavailability', *Journal of the American Dietetic Association*, vol 88, pp786–790

Munasinghe, M., Dasgupta, P., Southerton, D., Bows, A., McMeekin, A. and Walker, G. (2009) 'Consumers, business and climate change: Report by SCI with the CEO forum of companies', Sustainable Consumption Institute, University of Manchester, Manchester

National Food Administration (2009) 'Environmental-friendly food choices: Proposals notified to the EU', National Food Administration, Stockholm

Nestle, M. (2002) *Food Politics*, University of California Press, Berkeley, CA

Neumann, C. G., Murphy, S. P., Gewa, C., Grillenberger, M. and Bwibo, N. O. (2007) 'Meat supplementation improves growth, cognitive, and behavioral outcomes in Kenyan children', *Journal of Nutrition*, vol 137, pp1119–1123

Nierenberg, D. (2005) 'Happier meals – Rethinking the global meat industry', Worldwatch paper 171, Worldwatch Institute, Washington DC

OECD (2008) 'Sustainable development: Linking economy, society, environment', Organisation for Economic Co-operation and Development, Paris

Oliver, J. (2006) 'My manifesto for school dinners', available at www.jamieoliver.com/media/jo_sd_manifesto.pdf?phpMyAdmin=06af156b76166043e2845ee292db12ee, last accessed 25 May 2010

Pennington, H. C. (2009) 'The public inquiry into the September 2005 outbreak of e-coli 0157 in South Wales', Wales Assembly Government, Cardiff, available at http://wales.gov.uk/ecolidocs/3008707/reporten.pdf?skip=1&lang=en, last accessed 25 May 2010

Penningon, T. H. (2003) *When Food Kills*, Oxford University Press, Oxford

Pew Commission on Industrial Farm Animal Production (2008) 'Putting meat on the table: Industrial farm animal production in America', Pew Charitable Trusts and Johns Hopkins Bloomberg School of Public Health, Washington DC

Pew Oceans Commission (2003) 'America's living oceans: Charting a course for sea change', Pew Charitable Trusts, Washington DC

Phillips, L. P. O. W. M., Bridgeman, J. and Ferguson-Smith, M. (2000) 'The BSE inquiry: Report: Evidence and supporting papers of the inquiry into the emergence and identification of Bovine Spongiform Encephalopathy (BSE) and variant Creutzfeldt-Jakob Disease (vCJD) and the action taken in response to it up to 20 March 1996', The Stationery Office, London

Pollan, M. (2008) *In Defence of Food: The Myth of Nutrition and the Pleasures of Eating*, Allen Lane, London

Popkin, B. M. (1998) 'The nutrition transition and its health implications in lower income countries', *Public Health Nutrition*, vol 1, pp5–21

Popkin, B. M. (2009) 'Reducing meat consumption has multiple benefits for the world's health', *Archives of International Medicine*, vol 169, pp543–545

Popkin, B. M. and Gordon-Larsen, P. (2004) 'The nutrition transition: Worldwide obesity dynamics and their determinants', *International Journal of Obesity and Related Metabolic Disorders*, vol 28, ppS2–S9

Popkin, B. M. and Nielsen, S. J. (2003) 'The sweetening of the world's diet', *Obesity Research*, vol 11, pp1–8

Porritt, J. (2005) *Capitalism as if the World Matters*, Earthscan, London

Ravallion, M. (2004) 'Competing concepts of inequality in the globalization debate', World Bank Policy Research Working Paper Series, Washington DC

Rifkin, J. (1992) *Beyond Beef: The Rise and Fall of the Cattle Culture*, Dutton, New York

Rimas, A. and Fraser, E. D. G. (2009) *Beef: How Milk, Meat and Muscle Shaped the World*, Mainstream Publishing, Edinburgh

Rogers, B. (2004) *Beef and Liberty: Roast Beef, John Bull and the English Nation*, Vintage books, London

Rose, G. (1992) *The Strategy of Preventive Medicine*, Oxford University Press, Oxford

Royal Commission on Environmental Pollution (2004) 'Turning the tide: Addressing the impact of fishing on the marine environment', 25th report, Royal Commission on Environmental Pollution, London

Royal Society (2004) 'Royal Society infectious disease in livestock inquiry follow-up review', London

School Food Trust (2009a) www.schoolfoodtrust.org.uk/stacker_detail.asp?ContentId=435, last accessed 15 June 2009

School Food Trust (2009b) 'Meat products – Categorised and restricted', available at www.schoolfoodtrust.org.uk/stacker_detail.asp?ContentId=437, last accessed 10 December 2009

Sinclair, U. (1985 [1906]) *The Jungle*, Penguin, Harmondsworth

Singer, P. (1975) *Animal Liberation: A New Ethics for our Treatment of Animals*, Random House, New York

Singer, P. (2009) *Acting Now to End World Poverty*, Picador, London

Singer, P. and Mason, J. (2006) *Eating*, Arrow, London

Sinha, R., Cross, A. J., Graubard, B. I., Leitzmann, M. F. and Schatzkin, A. (2009) 'Meat intake and mortality', *Archives of Internal Medicine*, vol 169, pp562–571

Smith, M. J. (1991) 'From policy community to issue network: Salmonella in eggs and the new politics of food', *Public Administration*, vol 69, pp235–255

Stanton, R. and Scrinis, G. (2005) 'Total wellbeing or too much meat?', *Australasian Science*, vol 26, pp37–38

Steinfeld, H., Gerber, P., Wassenaar, T., Castel, V., Rosales, M. and de Haan, C. (2006) 'Livestock's long shadow: Environmental issues and options', FAO, Rome

Stern, N. (2006) 'The Stern Review of the economics of climate change', final report, H.M. Treasury, London

Stern, N. (2009) 'Climate chief Lord Stern: Give up meat to save the planet', *The Times*, available at www.timesonline.co.uk/tol/news/environment/article6891362.ece, last accessed 25 May 2010

Sustainable Development Commission (2009) 'Setting the table: Advice to government on priority elements of sustainable diets', Sustainable Development Commission, London

Symons, M. (1998a) *One Continuous Picnic: A History of Eating in Australia*, Penguin, Adelaide

Symons, M. (1998b) *The Pudding that Took a Thousand Cooks: The Story of Cooking in Civilisation and Daily Life*, Viking, Australia

Tukker, A., Huppes, G., Guinée, J., Heijungs, R., de Koning, A., van Oers, L., Suh, S., Geerken, T., van Holderbeke, M., Jansen, B. and Nielsen, P. (2006) 'Environmental Impact of Products (EIPRO): Analysis of the life cycle environmental impacts related to the final consumption of the EU-25', EUR 22284 EN, European Commission Joint Research Centre, Brussels

van Zwanenberg, P. and Millstone, E. (2005) *BSE: Risk, Science, and Governance*, Oxford University Press, Oxford

WCRF/AICR (2007) 'Food, nutrition, physical activity and the prevention of cancer: A global perspective', World Cancer Research Fund/American Institute for Cancer Research, Washington DC

WCRF/AICR (2009) 'Policy and action for cancer prevention – Food, nutrition, and physical activity: A global perspective', World Cancer Research Fund/American Institute for Cancer Research, Washington DC

WWF–UK (2009) 'One planet food strategy 2009–2012', WWF UK, Godalming

17

Confronting Policy Dilemmas

Jonathon Porritt

Apart from much of today's idiotic 'conspicuous consumption', there are only two facets of current human behaviour that simply cannot be squared with any reasonable scenario of a genuinely sustainable future for human-kind as a whole. The first is the continuing growth in demand for air travel, and the second is the continuing growth in demand for meat consumption. However traumatic the transformation from a carbon-intensive way of life to an ultra-low carbon way of life may prove to be (and the longer we delay, the more traumatic it *will* be), there are no other *insuperable* barriers from either a technological or a socio-economic point of view.

Aviation's contribution to total greenhouse gas emissions is still relatively small (anywhere between 2 per cent and 4 per cent, depending on how the calculations are done). Livestock production, by contrast, contributes at least 18 per cent of total greenhouse gas emissions, though here again the figures are complex.

Interestingly, that 18 per cent is almost exactly the equivalent of total greenhouse gas emissions from continuing deforestation. After years of neglect, what is known as REDD (Reducing Emissions from Deforestation and Degradation) is now very high up the list of priorities for world leaders. Ever-larger sums of money are being talked about in order to get a series of global deals in place.

By contrast, policy makers' attention to the 18 per cent from meat eating is as close to zero as it is possible to get. But for how much longer can that deplorable state of affairs continue? As the data get more robust (as mapped out in Chapter 3 in this book), academics and NGOs are now beginning to flex their muscles in terms of forcing the issue into the public domain. In 2004, through its Eat Less Meat campaign, Compassion in World Farming was pretty much a lone voice advocating decisive interventions by govern-ment to restrict projected increases in per capita meat consumption. Today, it's far from being a lone voice. Indeed, the food retailers are all carrying out their own analysis of the issue, and more and more people inside govern-ments in the EU know that the meat consumption challenge is just going to grow and grow.

In a previous publication for Compassion in World Farming, I felt it important to state my own personal rationale as a consumer of meat:

> *I write these words as a meat-eater. I've never been a vegetarian, and as a prominent exponent of all things sustainable, have often been attacked by vegetarians for what they see as inconsistency at best and outright hypocrisy at worst. I don't see it that way, though I'm aware that my own personal response to this dilemma (which is to try and eat a lot less meat and buy almost all the meat we consume as a family from organic suppliers) won't work for most people for reasons of price, availability and so on.*
>
> *Whilst I will always continue to campaign actively to improve the welfare of farm animals, and to eliminate all forms of cruelty from the food chain, I'm reconciled – with those caveats – to the moral accept-ability of the human species using other animal species for their own benefit. By contrast, I'm far from reconciled to the grotesque misuse of the earth's resources that our current pattern of meat-eating demands.*
> (Compassion in World Farming, 2004)

Building on that rationale, I believe it is perfectly possible to produce meat on a strictly sustainable basis and according to the strictest ethical criteria. But the implication of taking both of those two concepts seriously (sustain-ability and ethics) is that the total amount of meat consumed in the world will be massively reduced.

Exactly the opposite is happening today. According to the Food and Agriculture Organization (FAO), total global meat consumption is projected to grow from 240 million tonnes today to 376 million tonnes in 2030 and 465 million tonnes in 2050. In 2008, sharp price rises in all the major food commodities caused severe food shortages in many parts of the world, with food riots in around 25 different countries. Doubling meat consumption over the next 40 years will require the growing of at least 1 billion tonnes of additional feed. Even if such growth were 'a good thing' (and this book has demonstrated just how dreadful a thing it would in fact be), it simply won't be physically possible. So why do we permit chronic 'impossibilism' of this kind to dominate our worldviews and our policy deliberations?

The power of industry

Many of the reasons for this deeply unsatisfactory state of affairs have surfaced in preceding chapters. Politicians are hugely reluctant to incur substantial political risk where the upside in addressing the challenge is not as yet (for most of them) as clear as it needs to be, let alone for most of their voters. The risk resides in a number of different areas. First, the industry itself has become increasingly concentrated. With fewer companies commanding an ever greater share of the marketplace, the food industry has simultaneously become increasingly powerful, particularly in the US. A loose coalition of industry trade groups deploys armies of lawyers and lobbyists to protect their own interests – and even challenged the supremely influential Oprah Winfrey when she started championing one of the industry's critics. US democracy is a strange beast, and the role of the pork barrel (and indeed the beef barrel and the chicken barrel) remains disturbingly present.

Second, there will undoubtedly be substantial consumer resistance to any concerted measures to reduce average meat consumption. This may be an uncomfortable truth to surface in a book of this kind, but the majority of people rather like their meat, and a substantial minority like it a lot. It has taken years to raise concerns about the health effects of excessive meat consumption, and education campaigns to date have had little effect. The average US citizen consumes 100kg of meat per person per year (ERS USDA, 2009) – four times as much as is recommended for dietary purposes. Astonishingly, that figure is still increasing by about half a kilo per person per year – and we're not far behind that rate of increase here in the EU either.

Third, there is a whole set of legitimate concerns about the potentially regressive impacts of policies designed to reduce meat consumption. 'Cheap meat' has been built up as one of the great achievements of the last two or three decades – and so it is, just so long as you avoid doing a more detailed cost–benefit analysis. Meat is now so cheap that in both the US and (to a lesser extent) the EU it has somehow become 'culturally embedded', the default dietary option as much, if not more, for the poor than for the rich. China and other emerging economies are following rapidly in those ill-advised footsteps.

Lastly, there's now an important equity issue in terms of food as a globally traded commodity. The huge expansion in meat consumption in the rich world has opened up unprecedented opportunities for countries like Brazil and Argentina to provide the feedstuffs on which the livestock industry depends – soy, wheat and so on. Within the next decade, according to the Food and Agricultural Organization, the amount of meat that Brazil exports will be larger than that of the United States, Canada, Argentina and Australia combined (OECD/FAO, 2007).

Against that backdrop, politicians' finely honed survival skills are inevitably going to predispose them to inertia rather than to the kind of leadership that the evidence now demands. My nine years as Chairman of the Sustainable Development Commission taught me that ministers' enthusiastic advocacy of 'evidence-based policy making' pretty quickly evaporated once that evidence turned out to be either inconvenient or totally unpalatable. So even if Sir John Beddington, the UK's current Chief Scientific Advisor, were to take to ministers the kind of comprehensive dossier of evidence to be found in this collection, he would still get very short shrift.

Eventually, however, evidence wins the day. You just can't keep a good body of facts down – it will out. But even that, of itself, is not necessarily enough to ensure the kind of policy making process required for such a deeply controversial area. Seduced though I would happily be by Colin Tudge's eloquent account of how increased meat consumption has *not*, historically, been a principal indicator of status and affluence, I fear that history is against him on this one. As societies work their way out of poverty, meat consumption correlates with increased income in almost all circumstances where religious constraints (as with Hinduism) and cultural norms (as in Buddhist cultures) do not apply. I was delighted to hear from Colin that the Emperors of China infinitely preferred the skin of the duck to the flesh of the duck, but you can guarantee that hundreds of millions of their subjects would have loved to dine on duck every day. And right now in China, that's pretty much what they're doing, where the correlation between increased income (and the status that brings) and increased meat consumption could not possibly be clearer.

Criteria for successful policy making

In such circumstances, policy makers have to keep the objective simple: seek to reduce per capita meat consumption. Professor McMichael (see Chapter 11) has suggested that the appropriate threshold would be to get consumption down to no more than 90g a day, of which no more than 50g must be from red meat. This is against a 2005 baseline of around 300g a day, which on current projections would keep on gently rising rather than declining.

Personally, I'm not sure that the 90g threshold is sufficient, even from a climate change perspective, as this will do no more than stabilize the emissions associated with meat consumption at the 2005 baseline. The likelihood is that meat consumption, as a generic human activity, will have to take a bigger share than that if we're to get anywhere close to the 80 per cent cut in greenhouse gases required by 2050. But no need to scare the horses at this stage, let alone eat them, so we'll go with 90g. (I'm going to use the UK here as a proxy for all countries with unsustainably meat-intensive diets.)

If I was the Permanent Secretary in the Department for Environment, Food and Rural Affairs, listening to the Chief Scientific Advisor and realizing that this could no longer be avoided, I would recommend to the prime minister of the day that he/she should promptly set up an Advisory Council to determine 'how best to get from 300g to 90g a day within the next ten years'. Such a Council should include no representatives of the different vegetarian/vegan advocacy groups, and no representatives of the meat industry or the food industry more generally. It should be made up of people with impeccable scientific and non-scientific credentials, to ensure that their resulting conclusions/recommendations would give plenty of 'cover' for whichever ministers happened to be in the firing line when the going gets tough. Let's call it the '90g Advisory Council'.

They will have a lot of material to play with. The compelling truth is that there can be few theoretical 'single policy objectives' (in other words, reduce per capita meat consumption to 90g per person per day) for which there are more supporting evidence bases: health, food hygiene, animal welfare, climate change, wider environmental impacts and equity issues, all have a crucial part to play. Any one of these might (and, from an ethical point of view, *should*) suffice on its own – and the combination should surely be overwhelming.

The first thing to do would be to develop some kind of policy matrix. Such a matrix would need to assess the value of different policy interventions against a set of overarching criteria: *effectiveness* (just how substantive is the contribution any policy might make to the 90g objective?); *practicality* (just how easy will it be to implement and enforce any particular policy?); *acceptability* (just how intense are the objections to any particular policy likely to be?); and *desirability* (just how important would any policy intervention be, in the best of all possible worlds, from an ethical or values-based perspective?).

It is clear, as far as Kate Rawles is concerned (see Chapter 13 in this book), that once the science has been 'signed off', that any such policy making process should be *primarily* values-based, rather than geared to effectiveness and practicality, let alone public acceptability. And she is, of course, right. But world-weary Permanent Secretaries would dutifully point out that this is not 'the best of all possible worlds' and that the best of policies can sometimes be derailed by a combination of vested interests, ignorance and crude political prejudices. For instance, the panic-induced withdrawal back in 2000 of the government's admirable fuel-tax escalator, in the face of incensed lorry drivers, aggrieved farmers and a motley crew of petrolheads, was not only a massive blow to a still relatively new Labour Government, but basically kiboshed any interest that Treasury might have had in the whole question of eco-taxes and how best to effect behaviour change through fiscal instruments of this kind.

To reduce the risks of any such debacle on the reduced meat consumption front, the 90g Advisory Council would need, quite ruthlessly, to assess the correlation between any proposed policy instruments and people's *raw self-interest*. Though Colin Tudge is right to point out that the vast majority of people do indeed demonstrate significant 'natural sympathy' (to retrieve the long-overlooked phrase that Adam Smith used regularly to explain what would actually make the 'invisible hand' of the market work), that sympathy is never quite as powerful as the sympathy that people have for their own interests and the interests of those nearest to them.

Working with self-interest

So how would the hierarchy of self-interest look as the 90g Advisory Council settled into its first few months of work? They'd probably argue like hell about it, but having no particular vested interests to deal with, I suspect they'd come up with something like the following:

1 health;
2 climate change;
3 wider environmental impacts;
4 animal welfare;
5 equity/fairness issues.

I appreciate, from an ethical point of view, that such a hierarchy is probably deeply offensive to animal welfare campaigners. But how else can we account for the grindingly slow process of change over the last few decades to establish the bare minimum of 'decency' standards for animals kept in captivity? Very few people condone suffering or cruelty to animals when confronted by it, but not enough of them feel sufficiently concerned to have encouraged politicians to have raised standards faster and more comprehensively.

But they *do* care about their own health, and with a serious education campaign (think tobacco as the analogy here, or possibly alcohol consumption), there's no reason why more people would not start to think much more seriously about the impact of their personal meat consumption on their own health. It's just that we haven't had a serious education campaign as yet.

And the rationale really couldn't be clearer. As spelled out in Part 3, there's now a particular emphasis on obesity. The UK Government is already very focused on a growing crisis. Both the Foresight's 2007 Report on Obesity and the Cabinet Office's 2008 'Food matters' report demonstrated just how serious a policy challenge this has now become. The UK already has the highest rate of childhood obesity in the EU. The Foresight report predicted that 40 per cent of UK citizens will be obese by 2025 and 60 per cent by 2050 on current projections (Foresight, 2007).

In the US, it's even worse: the US Centers for Disease Control estimates that there are already 112,000 premature deaths every year arising from complications from obesity and related problems, amounting to around $75 billion in medical costs – that's nearly 5 per cent of total health care costs in the US (US Centers for Disease Control, 2006). Food-related ill-health cost the NHS an estimated £7.7 billion in 2007, 9 per cent of its total budget (Cabinet Office, 2008).

Tentatively, a few commentators *are* beginning to make some direct connections between obesity and climate change – not least through the compelling neologism of 'globesity'! The World Health Organization published some data in 2009 showing that each overweight person causes an additional 1 tonne of CO_2 to be emitted every year. With 1 billion people judged to be overweight around the world – of whom at least 300 million are obese – that's an additional 1 billion tonnes (Edwards and Roberts, 2009).

This is still very controversial territory. But on accelerating climate change in general, things are really starting to change. For more and more people, this is moving out of the 'possibly, but not in my lifetime' category of concern to the 'blimey, that doesn't sound good' category. Whereas the rhetoric was once geared almost entirely to appeals on behalf of 'our children and our children's children' (eloquently invoking the spirit of inter-generational justice), now it's all about *us*, in our lives, in our own backyards. Such is the measure of urgency that our growing scientific knowledge of climate change brings with it.

The chapter by Arjen Hoekstra on 'The Water Footprint of Animal Products' makes an extraordinarily convincing case as to this particular environmental issue. With delicious irony, one of the best overall summaries of these cumulative impacts can be found in the FAO's 2006 Report 'Livestock's long shadow' (Steinfeld et al, 2006). Unfortunately, this hasn't persuaded the FAO to start recommending to governments that they should start thinking about appropriate policy interventions – but, remember, 'you can't keep a good body of facts down'.

As to animal welfare, I think it's important to subdivide this into 'impacts on human health' and 'impacts on the animals themselves'. Quite frankly, having taken in the full analysis of Michael Greger's article (Chapter 10), it's perfectly clear that potential impacts on human health could go straight to the top of the 90g Advisory Council's hierarchy at any point in time. As we contemplate the full extent of a potential 'catastrophic storm of micro-bial effects', either directly or indirectly associated with intensive livestock production systems, it wouldn't take much to spill public opinion over from 'don't like the sound of that' to 'no more intensive systems of that kind *now* – and we don't care how much more our meat will cost'.

There's one particular aspect of this that is particularly potent: the continuing, widespread use of antibiotics in livestock production systems on a sub-therapeutic or even prophylactic basis. As our knowledge of antibiotic resistance in humans increases all the time, and medical concerns about different multi-drug resistant bacteria become ever more acute, pressure on ministers to regulate and even ban such practices is growing all the time. It is, to be sure, a rather indirect route to our 90g objective (banning the sub-therapeutic use of antibiotics would substantially raise costs of production, which would raise costs through the supply chain, which may contribute eventually to people turning away from meat and meat-based products), but it would be highly influential in terms of subtly transforming the whole cheap meat debate.

On the straight animal welfare issues themselves, I only wish that what Andy Butterworth describes in Chapter 8 as 'the margin for care' for animals held in captivity mattered far more to far more people. I'm persuaded this *will* happen in time, and the indefatigable efforts of animal welfare organizations the world over will in the meantime keep chipping away at raising standards and eliminating some of the cruellest practices. But I would be very surprised if the 90g Advisory Committee suggested to ministers that they should spearhead their new policy interventions primarily on the basis of improved welfare standards.

And that's also why concerns about the impact of *our* meat consumption on the lives of some of the world's poorest and hungriest people still come at the bottom of the list – despite the fact that in this age of cornucopian plenty there are still around 35 million hunger-related deaths a year. There is of course the most powerful moral argument (spelled out a number of times in this book) that the grain and other feedstuffs that currently feed animals should instead directly feed people. As Colin Tudge points out, by 2050 (on current projections), the livestock we eat will then be consuming the same amount of food as would be sufficient to feed 4 billion people directly, in effect increasing total human numbers at that time from an estimated 9.5 billion in 2050 to 13.5 billion.

But I fear that's not the way most people see things. Even if they did see it as a straight 'either/or' debate of this kind, we're then deep into people's complicated attitudes towards developing countries, food aid, corruption and so on.

Policy interventions

Armed with its Machiavellian matrix, our intrepid 90g Advisory Committee would then move to the consideration of specific policy interventions available to governments, all the way along a conventional policy spectrum with voluntarism at one end to outright interdiction at the other. Given that our objective is 90g a day rather than 'no meat at all', we need not concern ourselves any longer with a total ban. At the voluntary end of things, this is obviously the relatively easy place to start, with a commitment to a substantial and ongoing public education campaign to enable citizens to understand the full personal and societal implications of today's meat-intensive diets. 'Meat-free Mondays' and other 'Eat Less' campaigns are early signals of just how big this could be.

If past dietary challenges (on salt, for instance, or fat or artificial colourings) are anything to go by, the government will seek to shift as much of the burden involved in such a campaign onto our food retailers and (of increasing importance today) the catering industry. Portion size is already part of many retailers' 'healthy meals initiatives'. Alternatives to meat could be marketed much more aggressively, and there's considerable scope for further innovation here. For an industry that brings out more than 19,000 new products *every year*, the opportunities for a whole new family of 'I can't believe it's not meat!' products must be enormous!

The next level in is for any government to take responsibility for its own actions and for its use of public money. Throughout the world, billions of dollars of subsidy to farmers directly and indirectly promote increased meat consumption. An end to all such perverse subsidies (let alone to specific government-funded meat-promotion campaigns!), would have a huge impact on the economics of meat production. What is the logic of using taxpayers' money to promote patterns of behaviour that translate directly into an increased burden for the taxpayer in terms of health, environmental pollution, disease, climate change and so on? The case against production-related subsidies within the CAP is of course one that has been diligently pursued by the UK for many years. With the combined firepower of the health lobby and the climate change lobby weighing in behind that campaign, things could just move a little faster from now on.

Closer to home, the government would need to get its own food procurement policies sorted out. In its new climate change strategy, the National Health Service (NHS) has started to think through the implications of all this regarding hospital food:

> *The actions needed to develop a more sustainable food system in the NHS, whilst maintaining nutritional value, include the use of seasonally adjusted menus, increased use of sustainably sourced fish and a reduction in the reliance on meat, dairy and eggs.* (National Health Service Sustainable Development Unit, 2009)

And that would be just the start. Predominantly meat-free school dinners; reduced meat consumption in prisons; 100 per cent meat-free government canteens: there is absolutely no reason why this should not be taken completely for granted in ten years time.

Next up would be regulatory interventions – on pollution controls, higher welfare standards, a ban on the use of sub-therapeutic antibiotics, and so on. As already discussed, the effect of all this would be to raise the cost of meat for consumers – in effect to start consigning the concept of 'cheap meat' to the dustbin of history. But it's at this point that a couple of 'wicked issues' raise their ugly heads. First, how could we do this in the UK without further disadvantaging our own farmers, at the hands of intensive livestock operations free to carry on in the same old cruel, polluting and environmentally unsustainable way that we had just put behind us? This would be a disaster. Indeed, our farmers should in every respect be incentivized to be the pioneers in the transition to genuinely sustainable and ethical meat production systems.

Unfortunately, under World Trade Organization (WTO) rules, that's not possible. Just as we can't do a lot of other sustainable and ethical things under the wretched WTO rules. So how far might we here in the UK be prepared to go to reassert our right to pursue our own health, environmental and welfare interests?

The second wicked issue is, of course, social justice. Rationing meat consumption by price is a deeply regressive approach. For instance, David Pimentel of Cornell University has for a long time advocated some kind of 'sustainability tax' that would be applied to all foods on the basis of the external costs involved in bringing them to the market (Pimental and Pimental, 2008). Meat and dairy products would be taxed most heavily, whilst legumes, fresh fruit and veg wouldn't be taxed at all. But the problem is a simple one: this would still hit the poorest hardest.

Personal meat quotas

It's here that the overarching objective (90g a day per person, of which no more than 50g must be red meat) begins to bite. After all the regulation and the education and the incentivization, there are only two ways of further reducing demand. First, by raising prices until the inherent inelasticity in this particular market disappears. But meat is now so cheap that prices would have to rise very substantially indeed for that to happen, and the regressive impacts would be very severe. Second, governments could move to set 'personal meat quotas' along the lines of the much-discussed personal carbon quotas proposed as a way of reducing individual carbon footprints.

Were that to be the recommendation of the 90g Advisory Committee, the response would undoubtedly be something along the lines of 'are you all out of your tiny minds?!'. But science and logic are great bulwarks against accusations of insanity. And let's just remember the science behind this, just from the perspective of what is and what isn't physically possible. In his admirable book *The End of Food*, Paul Roberts puts it like this:

Under any model for a future food system that is both sustainable and equitable, the meat-rich diets of the West and especially the United States, simply don't work on a global scale. If the American level of meat consumption – about 217lbs per person per year – were suddenly replicated world wide, our total global grain harvest would support just 2.6 billion people – or less than 40% of the existing population.

Even if we use a more modest level of Western meat consumption such as that of Italy, where per capita meat consumption is about 80% of the United States, world grain supplies would still be adequate for just 5 billion people. In fact, according to the Earth Policy Institute, it's only when the world adopts an Indian level of meat consumption – that is, around 12lbs of meat per person per year – that current global grain supplies would be adequate to feed 9.5 billion people. (Roberts, 2008)

Which sets that kind of incontrovertible *physical reality* at odds with today's *political reality*. But if the science is robust then and the imperative is as strong as more and more people believe it to be, then Personal Meat Quotas (PMQs) would provide an elegant and effective policy intervention. Every citizen would receive the same quota, eliminating any regressive effects; it would be set on a sliding scale, starting with current consumption levels for Year 1, reducing down to 90g in Year 10, allowing people plenty of time to adapt; quotas could even be traded to begin with, although any such resale market would obviously favour rich carnivores over poor carnivores; and the whole thing would be done electronically, via a smartcard from which deductions are made automatically in the butchers, supermarket or restaurant.

Scary prospect for our politicians? Of course. For people who see cheap meat as part of their latter-day birthright, such a proposal is a total anathema. For people like Colin Tudge (whose simple maxim is 'plenty of plants, not much meat and maximum variety') it's just common sense based on a sound understanding of the role that meat has played in our lives over the entire evolution of the human species. For those somewhere in the middle, their first response is quite likely to be one of resistance. However logical such an approach might be, it would still be seen to be non-viable. Psychologically, there doesn't seem to be that much difference between PMQs and the kind of meat rationing we had back in the Second World War.

And that's true! But that's exactly the point. Very few people have yet grasped the fact that combating the threat of accelerating climate change will indeed entail going onto some kind of 'war footing'. As I said before, most of the changes that need to happen could be driven through by technological innovation, radical fiscal shift and integrated policy interventions that will make low-carbon living relatively pain-free and positively benign from all sorts of environmental, health and well-being perspectives.

But aviation and meat consumption are not amenable, I fear, to that kind of relatively conventional policy portfolio. Hence the need for sterner measures!

References

Cabinet Office (2008) 'Food matters: Towards a strategy for the 21st century', Strategy Unit, Cabinet Office, UK

Compassion in World Farming Trust (2004) 'The global benefits of eating less meat', available at www.ciwf.org/globalwarning, last accessed 23 December 2009

Edwards, P. and Roberts, I. (2009) 'Population adiposity and climate change', *International Journal of Epidemiology*, vol 38, no 4, pp1137–1140

ERS USDA (2009) 'Agricultural baseline projections: U.S. Livestock, 2009–2018', available at www.ers.usda.gov/Briefing/baseline/livestock.htm, last accessed 10 February 2010

Foresight (2007) 'Tackling obesities: Future choices', Cabinet Office, Chief Scientific Advisor's Foresight Programme, available at www.foresight.gov.uk/OurWork/Active-Projects/Obesity/KeyInfo/Index.asp, last accessed 25 May 2010

National Health Service Sustainable Development Unit (2009) 'Saving carbon, improving health', National Health Service Sustainable Development Unit, Cambridge

OECD/FAO (2007) 'Agricultural Outlook 2007–2016', available at www.oecd.org/dataoecd/6/10/38893266.pdf, last accessed 21 December 2009

Pimentel, D. and Pimentel, M. (2008) *Food, Energy and Society*, CRC Press, Boca Raton, FL

Roberts, P. (2008) *The End of Food*, Bloomsbury Publishing PLC, London

Steinfeld, H., Gerber, P., Wassenaar, T., Castel, V., Rosales, M. and de Haan, C. (2006) 'Livestock's long shadow: Environmental issues and options', FAO, Rome

US Centers for Disease Control (2006) 'Overweight and obesity: Economic consequences', March 2006, US Centers for Disease Control, Washington DC

Index